Challe SCIENCE PUZZLES

Erwin Brecher & Mike Gerrard

Sterling Publishing Co., Inc.
New York

Books by Erwin Brecher
Lateral Logic Puzzles
Surprising Science Puzzles
The IQ Booster
World Class Puzzles
Challenging Science Puzzles

Edited by Claire Bazinet

Library of Congress Cataloging-in-Publication Data

Brecher, Erwin.
    Challenging science puzzles / Erwin Brecher & Mike Gerrard.
        p.   cm.
    Includes index.
    ISBN 0-8069-9610-2
    1. Physics—Miscellanea. I. Gerrard, Mike. II. Title.
QC75.B74   1997
530—dc21   96-49194
                                                           CIP

10 9 8 7 6 5 4 3 2

Published by Sterling Publishing Company, Inc.
387 Park Avenue South, New York, N.Y. 10016
© 1997 by Erwin Brecher and Mike Gerrard
Distributed in Canada by Sterling Publishing
℅ Canadian Manda Group, One Atlantic Avenue, Suite 105
Toronto, Ontario, Canada M6K 3E7
Distributed in Great Britain and Europe by Cassell PLC
Wellington House, 125 Strand, London WC2R 0BB, England
Distributed in Australia by Capricorn Link (Australia) Pty Ltd.
P.O. Box 6651, Baulkham Hills, Business Centre, NSW 2153, Australia
Manufactured in the United States of America
All rights reserved

Sterling ISBN 0-8069-9610-2

CONTENTS

Acknowledgments

The laws of nature are immutable. They are, however, not recorded in any statute books that might be available to humanity to study the rules that govern our cosmos. So, in a laborious process going back to the dawn of history, scientific man has tried to fathom the mysteries of the universe. These collective efforts, spanning millennia, provide a rich source of challenging conundrums.

In compiling *Challenging Science Puzzles*, we enjoyed permission from other authors to use their material, including Scot Morris, Stephen Barr, and the *Encyclopedia Britannica*. Thanks should go to Sarah Gerrard for her patient support, and to Bob Anthony on whom we tested some of our ideas.

Logistic support was provided by: Kate Furphy, who dealt most efficiently with the manuscript and numerous revisions and corrections; Derek Bradley, who edited the manuscript; and Stuart Brown, who was responsible for the illustrations.

INTRODUCTION

"All men by nature desire knowledge"
—Aristotle (384–322 BC)

This book is for the curious.

We are surrounded by phenomena often dressed in everyday clothes which are rarely recognized as such until, at odd moments, we stop to ask ourselves: What arcane rules make the physical world behave as it does?

Whereas philosophy is concerned with the "why" in probing the innermost corners of spiritual existence, science explores the "how" by processes of inductive and deductive reasoning. The former process is an exercise in arriving at general hypotheses and theories from specific observations and experiments, while the latter reverses the procedure by starting with the theory to show that it will satisfactorily explain experimental results. This dichotomous approach is at the core of the scientific method, which was first articulated during the Renaissance but which in fact has been used throughout human history.

More often than not we take the workings of the physical world around us for granted without realizing the profound effect these, many apparently mundane, phenomena have on our understanding of the universe.

Sir Isaac Newton (1643–1727), one of the great scientists of all time, discovered the principles of universal gravitation following the fall of an apple. The story might be anecdotal, but Newton's achievements are monumental.

The time is New Year's Eve 1670. Picture yourself sitting at home and seeing an object fall. Would you have given it a second thought? Probably not. But if you had, you would most likely have said to yourself: "What else could it do but fall to the floor?" In other words, you would have considered it axiomatic that everything falls down and not up. However there are no axioms in physics; that is to say, there are no basic principles that are assumed to be true without proof or inherent logic, as is frequently the case in geometry.

On the contrary, there is not a single aspect of our physical world which is not investigated, analyzed and evaluated by scientists who are determined to find an explanation for all phenomena, be they esoteric or mundane. The relevant literature is enormous, with some questions remaining unresolved while opinions on other theories differ.

We can't all be Newtons, but we can have an inquiring mind and derive a great deal of satisfaction from understanding what makes things "tick." In some cases we shall find the answer, using basic principles of physics and a modicum of problem-solving ability. With others the answer will make immediate sense, with a tinge of regret for not having thought of the explanation in the first place. Finally there will be quite a few where the solution will be a revelation.

In all cases, the intelligent reader will close this book feeling that his horizon has widened beyond expectation.

Erwin Brecher

I have always loved puzzles of all sorts, from crosswords to detective novels. There is something immensely satisfying about tracking down the solution from a few apparently unconnected bits of information. I think that is why I have always been fascinated by science. The scientific method of proposing hypotheses and then checking them out by experimentation is immensely attractive.

But solving a scientific conundrum has an additional

attraction that no thriller writer can hope to match. Having worked out that "the butler must have done it" is satisfying, but at the end of the day it is just a story. Solving a scientific puzzle also teaches you something about the way the universe works.

It must be one of the most exciting and creative of human activities to discover something absolutely new that no one has ever known before. The Einsteins, Newtons, Darwins, and so on, must all have shared an incredible sense of elation having made that final connection in their minds.

This book does not pretend to be in quite the same league, but it is hoped that you can experience a similar sensation. You may find an observation that you have never seen for yourself, or a question that you have never thought to ask.

The solution to it lies in your own head somewhere; all you need do is formulate your own thoughts to produce the solution. The process can be completely absorbing; it is where the stories of absent-minded professors originate. Archimedes was so involved with his own problem that he foolishly asked an invading Roman soldier to get out of the light, and got himself killed.

To get the most out of this book you are advised not to rush between the problems and the stated solutions; that will spoil the fun! Think about things. It may take several days or a little research to decide on what you think the answer is. Only then look at our solution. You may even disagree with what we say—your solution may be superior to ours. Happy solving!

Mike Gerrard

HEAT, LIGHT & SOUND

LED Lights

A light-emitting diode (LED) is a small solid-state device that gives off light when a current is passed through it. It is commonly used in clocks to give a clear red, or sometimes green, digital display.

There is a type of LED on the market that glows red when the current is passed one way through it, and green when the current flows the other way. An alternating current flows rapidly one way and then the other. What would this LED look like if it were connected to an alternating source?

Solution on page 55.

Black to the Future

Our modern world is full of light. Walk through any city center at night and you will see all sorts of colored lights, from white through every color in the rainbow.

But it occurs to me that there is one color that is omitted, and that is black. This seems a great pity, because I could think of many uses. For example, one could use a black bulb in a photographic darkroom, avoiding the necessity of costly blackout material. During a war, searchlights were used to illuminate bombers at night and also to dazzle the pilots. A black searchlight could be used during the day to plunge the flyers into darkness.

Anything wrong with this scheme?

Solution on page 56.

Ringing Tone

I was recently putting in some fencing in a field. While knocking in the fence posts, I noticed differences in the sound. In the center of the field I was a long way from anything, and the sound from the hammering of the posts was dull and flat. I concluded that this was because there was nothing near to reflect the sound. In another part of the field I could hear a distinct echo from a nearby building.

However, in a third part of the field I could hear a distinct ringing tone. Can you think what might have been the cause of the ringing sound?

Solution on page 61.

Lo-Fi?

When setting up my new hi-fi system, I noticed that the instructions were very emphatic about connecting the wires to both speakers in the same way. In fact, the wires were color coded to try to make it difficult to get wrong.

What would happen if the loudspeakers were incorrectly connected?

Solution on page 57.

Sound Reasoning

Most people know that light, like sound, is a wave form. Yet light and sound behave very differently. Can you explain:

1. Why it is you can hear around corners, but not see around corners?

2. Why you can use sound to cancel out other sounds, but cannot do the same with light?

Solution on page 58.

9

Frequent Changing

On a recent drive around the country, I was listening to a very interesting program on the car radio. As I moved away from a transmitter, the reception got progressively worse, so I had to retune the radio to pick up the same station on a nearer transmitter. I had to repeat this process several times during the journey. This really affected my enjoyment of the program. Why do they not ensure that all transmitters broadcasting the same station use the same frequency?

Solution on page 57.

Faster Than Light?

The other day, I was trying to explain relativity theory to a friend. Einstein had predicted, and experiments had confirmed, that as objects moved faster they also increased in mass. At the speed of light, any object would have infinite mass and so would need infinite energy to make it go faster.

"This means that there is a maximum speed at which anything can travel, and only massless things like light can travel at that speed," I explained.

My friend is very bright, and after a little thought he came up with three situations that seemed to contradict Einstein. Is any of them valid?

1. The poles outside barber shops have helical stripes painted round them. If the pole is rotated, its stripes are seen to travel along it. If the rotation is fast enough the stripes should travel faster than light.

2. If you shine a flashlight on a wall, you will see a spot of light. You can make this spot move along the wall by rotating the flashlight. The speed of the spot depends on the distance from the wall and the speed of rotation of the flashlight. If you shine a laser on a target some miles away,

the spot of light formed could be made to move faster than the speed of light by rotating the laser.

3. Very small particles, such as neutrons, can penetrate matter very easily. Imagine such a particle, traveling at nearly the speed of light, entering a block of glass. The particle is so small that it will not be impeded by the glass, but light is slowed down to about two-thirds of its speed in air. Therefore, the particle would be traveling faster in the glass than the light.

Solution on page 65.

The Light Fantastic

Imagine two blacked-out rooms bathed in yellow light. A piece of white paper examined in both rooms appears to be exactly the same shade of yellow. Another piece of paper appears black in one of the rooms and striped red and green in the other. How could this be?

Solution on page 61.

The Rainbow

Rainbows, an enchanting spectacle of nature, pose a number of baffling questions. We know they are caused by drops of water falling through the air, refracting sunlight in such a way that it creates a curve of light exhibiting all colors of the spectrum in their natural order. Now answer the following:

1. Why is it you do not always see a rainbow when it rains while the sun shines?

2. Can the moon also produce a rainbow?

3. Even without rain you can, at times, see a rainbow if you look across a lawn early in the morning. Why?

Solution on page 62.

Getting Cold Feet

The other night I could not sleep and was feeling hungry. Deciding to raid the refrigerator, I stepped, barefoot, from the carpeted hall into the tiled kitchen and found that the floor was extremely cold. The door between the hall and the kitchen had been open all the time. So why was the kitchen floor so much colder than the hall floor?

Solution on page 59.

Car Headlights

Many night driving accidents are caused by the glare of headlights from approaching cars. My friend Jonathan, who fancies himself an inventor, came up with an idea offering an effective and inexpensive solution to the problem.

"Let us introduce polarizing filters in front of headlights," he said, "to polarize light horizontally, which would absorb all photons whose electric vectors are vertical. Conversely, let us use windshields with a filter turned 90° to the first, absorbing light emitted by the headlights." On the face of it, this would be the perfect solution, as the light from the approaching car would be blocked out, while all other objects would be visible.

Would this idea work? If so, why hasn't it been adopted?

Solution on page 56.

Vision and Sound

Why are humans endowed with two eyes and two ears? Is the second organ just backup equipment or a spare, or has it another function?

Solution on page 63.

The Toy Boat

When, as a boy, I built my own little steamboat to operate in the bathtub, I was thrilled. It was my own design, using odd bits of metal plus a little fatherly help, and looked something like this:

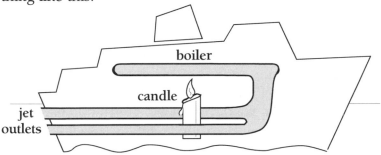

The principle was simple. The candle heated the water in the boiler and the resulting steam forced the water through the outlets as jets, propelling the boat forward. To my surprise I noticed that, from time to time, the outward jets stopped and instead water was sucked in through the tubes. I did not understand why this should be and, more surprising, why the boat did not then reverse direction and move backwards, as a basic law of physics would suggest (Newton's Third Law of Motion: action = reaction).

Can you explain?

Solution on page 62.

Breath on Your Hand

If you breathe gently on the back of your hand it feels warm. If you purse your lips and blow hard on the back of your hand it feels cold. Surely the temperature of one's breath is the same on both occasions, so why does it feel different?

Solution on page 59.

The Foehn

It has long been accepted that weather has a profound effect on the physical well-being and even the mood of most people. One of the well-known natural phenomena is a warm dry wind blowing down from high mountains into valleys. In German-speaking countries this wind is called *foehn*, though it has many other names with one thing in common: dry, warm and unpleasant. *Foehn* or its equivalent is held responsible for severe headaches and, in the extreme, for criminal behavior in some of those affected by it.

How can a warm wind come down from a cold mountain, reach speeds of close to 75 mph, and have such a physiological impact on many people?

Solution on page 57.

The Drinking Bird

Many toys and promotional gadgets are based on physical phenomena, often not readily understood. The drinking bird is one of the most fascinating examples. It consists of a glass bird, standing in front of a container filled with water.

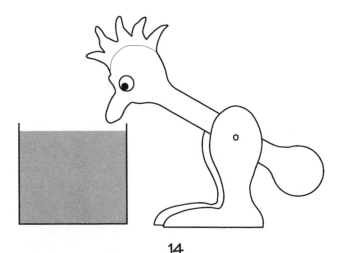

The bird appears to be thirsty at regular intervals and so it rocks forward and dips its head into the water. Having quenched its thirst, the bird rights itself, only to repeat the performance after a little while. All you have to do to get the operation going is to wet the felt that is covering the bird's head and beak, after which the bird continues to bob up and down without further assistance.

What makes it do this?

Solution on page 60.

The Safety Lamp

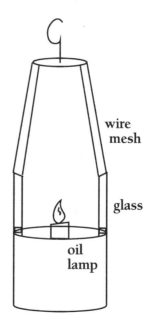

During the 18th century, miners were at extreme risk when exposed to explosive gases. It was in 1815 that Sir Humphrey Davy (1778–1829) invented a safety lamp to be used in coal mines. A fine metal screen in the form of a cylinder covered the open flame of the miner's oil lamp.

Did the screening prevent explosive gases from entering the safety lamp, or is there a different explanation for the lamp's effectiveness?

Solution on page 56.

Perpetual Motion I

To avoid misconception, it is important to understand the essence of perpetual motion and to realize that such a contraption would violate the laws of nature. Nonetheless, many such machines have been proposed throughout history. Illustrated is one ascribed to the Bishop of Chester (1674).

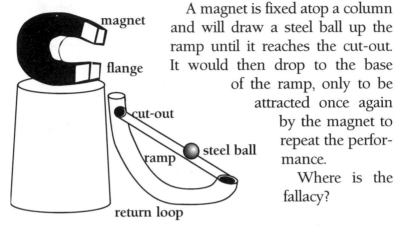

A magnet is fixed atop a column and will draw a steel ball up the ramp until it reaches the cut-out. It would then drop to the base of the ramp, only to be attracted once again by the magnet to repeat the performance.

Where is the fallacy?

Solution on page 58–59.

Upon Reflection

It is often said that mirrors reverse left–right but not up–down. Can you think how a single plane mirror can:
1. reverse up–down as well?
2. reverse up–left and down–right?
3. reverse up–down but not left–right?

Solution on page 66.

The Setting Sun

Sunsets after a blissful summer's day, with the beautiful variety of glorious colors in the fading daylight, are an uplifting experience and the stuff that poems are made of. It is almost blasphemous to try to explain the splendor of this phenomenon in terms of cold physics. Can you, if you tried?

Solution on page 56.

SPACE

The Fourth Dimension

Our physical world is said to be three-dimensional. While we can readily understand the meaning of zero, one, and two dimensions, the inadequacy of human imagination cannot normally cope with the existence of four-dimensional space. There are some who claim that they can, such as the nineteenth-century German physicist Hermann von Helmboltz, who maintained that the human brain was up to the task provided it was supplied with the necessary input data; but this transcendental feat seems beyond most of us. However, it is one thing to claim, and quite another to prove, that the human mind is capable of visualizing a fourth geometrical dimension.

Mathematics has no such difficulty. It accepts the existence of dimensions beyond the third; it does not stop at the fourth dimension, in fact, but considers the possibility of dimensions existing through to the nth dimension as a matter of logical progression. Many readers will consider this subject to be esoteric in the extreme and of interest only to scientists and philosophers. Yet the concept of a fourth dimension is the key to solving one of the most intriguing mysteries of our universe and its structure.

Is space infinite or finite? One is as difficult to contemplate as the other. Infinite means that no matter how far you travel in one direction there is still space in front of you, never to end. "Can't be," you say to yourself. "Everything has a limit." But what is the alternative? If space is finite, does it mean that if you keep traveling in a straight line in any direction you will come up against an impassable barrier, the end of space? Equally mind-boggling.

In the year 1922, the Russian physicist and mathematician Alexander Friedmann developed a concept of the universe based on certain assumptions which have been proven remarkably accurate by cosmological discoveries in the seventy-plus years since. Friedmann's model postulated the following characteristics of the universe:

1. All galaxies are moving away from each other.

2. The farther apart they are, the faster galaxies move away from each other, the relative rate of acceleration being in proportion to the distances between them.

3. There is no point in the universe that can be said to be the center of the expansion.

Many cosmologists have come to view our universe as a four-dimensional sphere (hypersphere) with a three-dimensional surface, having a circumference of the order of 100 billion light-years. One light-year is equivalent to 9,467,280,000,000 kilometers (5,917,050,000,000 miles). The suggested circumference of the universe is therefore approximately 9.467×1023 km.

According to this model, what we perceive as straight, parallel lines are in fact great circles intersecting at two points about 50 billion light-years distance, in each direction, on the hypersphere—in the same way that, on our globe, meridians meet at the poles. This model assumes a fourth dimension and suggests that the universe is finite, but without boundaries.

In my books, I try to prove that 4-D space is likely to exist. To achieve this, I use a speculative analogy. Imagine that somewhere in the universe there exists a 2-D planet —that is, a world whose inhabitants believe that theirs is a two-dimensional universe. How would we set about trying to prove to them that they were surrounded by three-dimensional space? Assume also the existence of a 2-D Friedmann who postulates that, as 3-D Friedmann does:

1. All galaxies are moving away from each other.

2. There is no point in the universe that can be said to be the center of the expansion.

Accept also that Friedmann's model has been proven "remarkably accurate" by cosmological discoveries. 2-D cosmologists will still need to reconcile the second postulate with 2-D geometry, say a flat sheet of expanding paper. No matter how hard one tries, there will always be a center of expansion, a point on the sheet that will remain stationary.

We, in our 3-D universe, know that the postulate can be accommodated if we assume that our 2-D friends are living on the surface of a sphere, say a balloon, which, when inflated, would indeed not have a center of expansion on its surface, while all galaxies would move away from each other. Their universe could then be said to be finite but without boundary.

It is only a small step to project this concept into our universe: that is to say that we, in our 3-D world, are on the surface of a 4-D sphere.

The question to the reader is this: Can you think of any other phenomenon or problem in the 2-D world, solvable only in the 3-D universe, which similarly exists in our 3-D environment? Could a fourth dimension provide the answer?

Solution on page 64.

In a Spin?

A long time ago, during a geography exam, I was having a real problem. I could not remember whether the Earth rotates east-to-west or west-to-east. I desperately tried to recall whether the sun rose in the east or west or American time was ahead of or behind European time, so as to work out the answer. I glanced anxiously at the clock as time went by. How do you think I solved the problem?

Solution on page 61.

Squaring the Circle?

Prolonged weightlessness can have unfortunate physiological effects on the human body and may prove a great problem on long space flights. It is well known that spinning the spacecraft can produce an effect very similar to gravity, so we have seen science fiction films where the astronauts live in circular living quarters which are rotating to provide apparently normal gravity.

However, as humans we are not used to living in round rooms. What would the effect be if the space explorers were put into a normal rectangular room rotating about a line drawn across the middle of the ceiling?

Solution on page 61.

One Side of the Moon

It is common knowledge that we on Earth see only one side of the moon. In other words, the moon rotates at a rate around its axis which synchronizes it with the Earth's rotation. Is this a coincidence?

Solution on page 65.

20

In the Dark?

Eclipses of the sun occur when the moon moves in front of the sun. The moon takes about four weeks to orbit the Earth. However, very few people have actually experienced a total eclipse of the sun.

Why is this? Why is there not a solar eclipse monthly?

Solution on page 59.

Information Relay

I have just watched a news item on television about the launch of the SOHO observatory satellite. There was a computer graphic showing that it was to be placed in an orbit halfway between the sun and the Earth, with exactly the same period of rotation as Earth. From that position it could observe the relatively nearby sun and relay information back to the Earth.

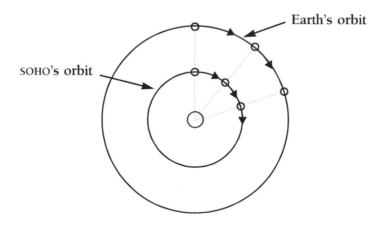

What is wrong with this explanation?

Solution on page 63.

Gravity

An astronaut awakes in an enclosed room where gravity is apparently normal. He realizes that there are three possibilities:

1. He is subjected to gravity.
2. The spacecraft is accelerating.
3. The spacecraft is creating its own "artificial gravity" by spinning.

Is there any experiment that the astronaut can do inside the room to discover which is the true situation?

Solution on page 62.

The Spy Satellite

After the Desert Storm war, the United States wanted to keep a close and constant watch over Iraq. To achieve this objective the U.S. directed spy satellites which, powered by solar cells and using long-range cameras, transmitted important information to the intelligence agency on Earth.

The ideal solution would have been a satellite hovering over Baghdad 24 hours a day. Instead, the U.S. used a number of them in orbit, changing place as in a relay race. Why?

Solution on page 63.

Pull of the Sun

Surely the sun's gravity pull on the moon is much larger than the Earth's. Why then does the sun not pull the moon away from its orbit around the Earth?

Solution on page 64–65.

Shape of the Flame

What would be the shape of a candle flame inside a spacecraft where there was normal air but no gravity?

Solution on page 62.

The Satellite

Most man-made satellites will eventually return to Earth, because their orbit is affected by the Earth's atmosphere which, though extremely thin, still exists very high up. However, the surprising effect of this air-drag is not, as one would expect, to slow the satellite down. On the contrary it will accelerate in its orbit.

How do you explain this phenomenon?

Solution on page 63.

A Moon Mystery

We all know that we, on Earth, never see the other side of the moon, and that moonshine is only reflected sunlight. However, when the moon is starting to wax and the new moon is only a narrow crescent, we believe that we can just about also see the remaining dark part of the moon.

Are we simply imagining it because we know it is there, or what?

Solution on page 69.

What Color Is the Sun?

This might seem incredibly easy to answer, but how many different answers can you think of to the question?

Solution on page 82.

Round or Flat?

If I were to ask you such a question now, referring to the shape of the Earth, you would hardly be amused. But transport yourself back in history a little more than 2,000 years and the question as to whether the Earth was a round sphere or a flat disc was by no means decided.

What evidence could you have mustered to decide one way or the other?

Solution on page 66.

The Olbers Paradox

Heinrich Wilhelm Matthaeus Olbers (1758–1840), a German astronomer, hypothesized that if the universe were infinite and contained an infinite number of stars, then the whole sky should be as bright as sunlight. After all, the observable universe alone indicates that the Milky Way is only one of several hundred thousand million galaxies, with each galaxy containing in turn some hundred thousand million stars. It seems therefore inconceivable that any line of sight could miss a star.

You can compare it with a blackboard on which you are asked to make a mere one million chalk marks. Surely the blackboard would appear white? Why then do we not see the sky brightly lit at night?

Solution on page 67

HOUSEHOLD PHYSICS

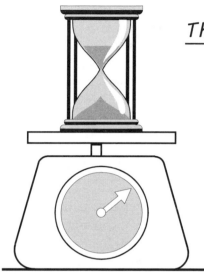

The Hourglass

You have an hourglass on the table in front of you, which has run its course. What happens if you turn it upside down? Will the weight of the hourglass be reduced while some of the sand is in free fall?

Solution on page 71.

Hotheaded

Having washed my hair the other night, I was trying to dry it as quickly as possible. I therefore had my hair drier on its highest setting. I noticed that the air still felt very cold on my head, but burned my ears. Why?

Solution on page 76.

The Toy Balloons

You have two identical balloons. Balloon "A" has been blown up to 10 cm. diameter and Balloon "B" to 20 cm.

You want to blow ten times into each balloon. Will it be easier to succeed with one balloon or the other, or will the necessary force be the same?

Solution on page 69.

So Near and Yet So Far

We are very used to the idea of telescopes and binoculars, but what are they doing? Do they make distant objects look closer or bigger?

What about a magnifying glass: does it make things *look* bigger, or is it doing something else?

Solution on page 67.

Insulated from the Truth

I was reading an advertisement the other day. It had been placed in a newspaper by a firm of energy consultants, and showed a diagram of an average house to illustrate the following figures: 50% heat loss through the walls, 30% through the doors and windows, and 20% through the roof and floors.

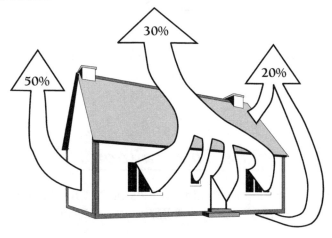

The firm claimed to be able to cut each of these heat losses by half. What is wrong with this claim?

Solution on page 66.

Glaringly Obvious?

Ordinary sunglasses work by absorbing some of the light that would otherwise be transmitted. Polarizing sunglasses work in a slightly different way: they will transmit only light that is vibrating vertically. As light reflected from wet roads, water, and so on tends to vibrate horizontally, this cuts down glare from such surfaces. If you are wearing a pair of polarizing sunglasses and hold up another polarized pair rotated through 90 degrees, the lenses appear black because no light can now pass through both sets of glasses.

Could you insert a third set of polarizing sunglasses between the first two in such a way as to permit some light to be transmitted?

Solution on page 68.

A Quiet Cup of Tea

There is nothing more cheerful than the sound of a kettle coming to the boil and the prospect of a refreshing drink to come. Have you noticed that the kettle goes strangely quiet just before the boil begins, and can you explain why?

Solution on page 69.

A Balancing Act

My kitchen scales are balanced when there is nothing on them, but I know that there is something wrong because I get different results depending on which side I put the object to be weighed. If an object balances with 100 grams when on one side, but with 144 grams on the other side, is it possible to determine the real weight?

Solution on page 68.

The Hot Water Tap

I could never understand why, whether at home or in a hotel, after turning on the hot water the flow of the stream first decreases, only to return to normal after about 30 seconds. Do you know what causes this?

Solution on page 69.

Raw or Cooked

You have an egg before you. How can you find out whether it is hard-boiled or raw—without, of course, opening it?

Solution on page 89.

The Expert Driver

You have driven a car for many years, and fancy yourself an expert. You are driving along a highway at 70 miles per hour. The weather is pleasant and you have not a care in the world. Suddenly you see a small animal trying to make it across the road. You cannot avoid it by swerving because there are cars on your left and right. You are lucky that there are no cars closely behind you because you are an animal lover and the only way to avoid killing the poor thing is to brake.

Will you slam on the brakes and lock them, or will you apply increasing pressure hoping you can stop just in time?

Solution on page 69.

Golf Balls

In the early days of golf the balls were smooth. Dimples were introduced later, after manufacturers claimed the dimpled variety traveled farther. Were they right, and if so why?

Solution on page 72.

Color of the Colorless

Clean air, water, and transparent glass are supposed to be colorless. Yet the sky is blue, a sheet of glass looked through sideways is green, and a mountain lake, pure as it is, appears blue.

Textbooks in physics will talk about "Rayleigh scattering," "electric field," and "electron oscillations," but there is a simpler, if unscientific, explanation. Do you know what it is?

Solution on page 70.

Cool It!

You are in a hermetically sealed, perfectly heat-insulated very warm room containing a large refrigerator. Can you, more or less permanently, reduce the room temperature?

Solution on page 71.

Soap Bubbles

One of the most beautiful sights in our physical world is that of soap bubbles showing all colors of the rainbow. Is the air pressure inside the bubble larger or smaller than, or equal to, the atmospheric pressure outside the bubble?

Logic will provide the answer.

Solution on page 70.

The Spray Gun

As children, it was fun to annoy grown-ups by spraying them with water, using a gadget as illustrated.

By blowing through the horizontal tube, you are forcing water up the conduit tube into a fine spray. The same principle is used in aerosols and paint sprayers. Why does water rise in the tube, against gravity?

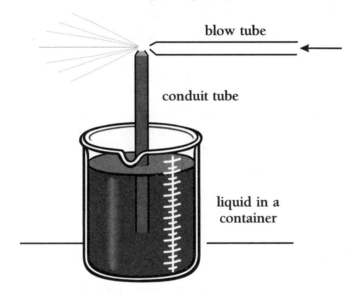

Solution on page 70.

The Bicycle Pump

Here is an easy one, to give your neurons a rest. When you pump up a bicycle tire by hand, the valve gets hot and so does the hand-pump on the downward stroke. If, however, you use a compressed-air bottle instead, the bottle cools.

Can you explain the different effects?

Solution on page 71.

What Makes Glue Stick?

You will probably say that this is easy to understand, but difficult to explain. If pressed further you might ascribe it to some chemical property of the adhesive. This is just begging the question, and furthermore is wrong. The answer is rather more complex. Have another guess.

Solution on page 72.

Insulating Metal Pipes

Here is a problem that might require a little more than elementary household physics.

Exposed hot-water and steam pipes are often insulated to reduce heat loss. What is surprising is that the "insulator" is often a better conductor of heat than air; for instance, cork is twice as good as air at conduction. Yet one would have thought that using such a material would be counterproductive. Why do we insulate pipes?

If the thermal conductivity of the insulation is so unimportant, why do we not just use thicker pipes? That is to say, pipes with the same internal diameter but greater external diameter.

Solution on page 74.

Grounding the Automobile

You will often see trucks and cars with a metal strip or chain hanging from the rear, being dragged along the ground. What is the purpose of this, and is it effective in what it is supposed to achieve?

Solution on page 73.

The Well-Trained Sunglasses

You will come across many puzzles in science to which no amount of reasoning will provide the answer if you do not happen to be a specialist in that particular field. Take, for instance, photosensitive sunglasses. It is not too far-fetched to recognize that sunlight can cause a physio-chemical reaction that will darken certain substances. After all, your skin tans on exposure to the sun. But tanning takes time. Furthermore, your skin does not turn pale again the moment you retreat into shade.

Can you at least speculate as to the principle that might be involved in the case of our well-trained sunglasses?

Solution on page 70.

Driving on Ice

Whether you are a good or bad driver makes all the difference in hazardous road conditions. Assume it is freezing and the road is covered with a layer of ice. Can you answer the following questions?

1. If you want to start the car and all you have to help you is one blanket, will you use it as underlay for the front or rear tires?

2. Do you put the car in low or high gear? And will you quickly accelerate or maintain low speed?

3. If you succeed in moving but the car starts to skid, will you steer into the skid or maintain direction?

Solution on page 89.

GENERAL
PHENOMENA

Water Level

There is a classic problem about throwing a brick into the water from a boat floating in a swimming pool. The question is whether the water level will fall, rise, or remain unchanged.

The story goes that several famous physicists, including Albert Einstein and Robert Oppenheimer, answered incorrectly. This story is surely anecdotal and possibly slanderous, unless both scientists were indisposed or playing truant when the Archimedes' Principle was explained at school.

While the boat/brick problem is well known, there is an interesting alternative problem. What happens to the water level if the boat leaks and slowly sinks? Does the water level change while it sinks or not? If it changes, does it rise or fall?

Solution on page 73.

Crafty Hover

A hydrogen balloon will rise into the air. Suppose that we wanted a balloon that neither rose nor sank, but hovered at a constant height. We will do this by attaching a length of string so its weight is just enough to balance the upward force. You could do this by starting with a piece of string which was too long and successively snipping bits off until it just balanced. The trouble is that if you cut off too much, you will have to start all over again. Is there a simple way to achieve the desired effect?

Solution on page 72.

Dropping a Bombshell

You may have seen films of planes dropping bombs to which small parachutes have been attached. Why is this done? Surely it is not to ensure that the bombs drop gently?

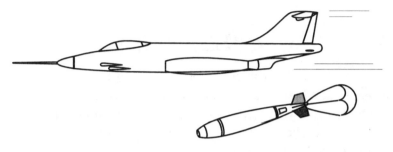

Solution on page 75.

Time to Leave?

In temperate zones, deciduous trees shed their leaves in the autumn. The trigger for this appears to be the shortening length of the days, even if the weather is still quite mild.

Surely it would be in the trees' interest to keep their leaves until the very last moment, in other words until the temperature fell to a dangerously low level. Can you find an explanation, considering that "Mother Nature" is an efficient professional.

Solution on page 67.

Long Live the Evolution

There are many diseases that are primarily associated with old age. As a species, will mankind ever evolve a resistance to such ailments.

Solution on page 75.

My Head in the Clouds

I was on a hiking holiday, when I noticed a peculiar phenomenon. At first, I could not work out what it was that "did not feel right." I was walking across a large plain. To the west was a range of mountains that stretched north to south as far as I could see. There was a strong wind blowing from the direction of the mountains.

I suddenly realized what seemed odd. It was the clouds. They were an unusual shape, forming long straight bands parallel to the mountains. But the really peculiar thing was that, despite the strong wind, they did not appear to be moving at all. Can you explain?

Solution on page 75.

Ring Around?

Suppose we created a ring made of some strong material all around the world supported by small pillars. Furthermore, supposing then, at a given signal, all the pillars are removed at the same time. Would the ring remain, apparently floating?

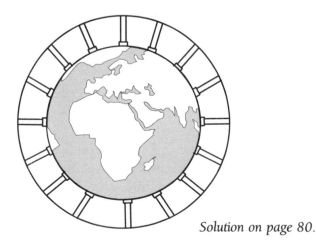

Solution on page 80.

Not the Half of It!

Suppose we have a number of flat objects, each with a different shape: square, circle, and so on. If we select any of those objects we could draw a bisector across it. I will define a bisector as a straight line that divides the object into two halves. Of course for any object we could draw an infinite number of bisectors. Would all the bisectors necessarily go through the same point for any specific object?

Solution on page 74.

Helicopter Flight

In airplanes, the lift is supplied by the wings while the thrust is supplied by the engines. Jet engines can make the plane move very fast indeed.

In helicopters, the lift is supplied by the rotors, which also provide the thrust by tilting the nose down. Could we get helicopters to go as fast as a plane if they were propelled by jet engines while the rotors were used solely to provide lift?

Solution on page 74.

Sheep May Safely Graze

There are many examples of two different species in an evolutionary race: for instance, cheetahs and gazelles each increasing their speed so that one can eat while the other can escape.

One might assume that similar competition exists between grass and grazing animals. Over the millennia, why has grass not evolved to be indigestible or poisonous to sheep, cows, rabbits, etc.?

Solution on page 78.

A Question of Stability

Scientists recognize that objects can have three types of stability:

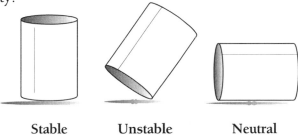

| Stable | Unstable | Neutral |

1. **stable**—tilt the object a bit and it returns to its original position, e.g., a cylinder sitting on its base. (Note that a cylinder has two stable positions because it has two bases.)

2. **unstable**—disturbing the object just a little makes the object move to a completely new position, e.g., a cylinder balanced on an edge. (Note that a cylinder has two unstable positions because it has two edges.)

3. **neutral**—rotate the object and it stays in its new position, e.g., a cylinder on its side. (Note that a cylinder has one neutral position because it has one side.)

We could call a cylinder a 2:2:1 object, because that is the ratio of its *stable:unstable:neutral* equilibria. Can you find examples of the following objects:

 a) 1:2:1 b) 0:0:1 c) 0:2:1 d) 1:1:0

Can you prove that 0:x:0 is impossible?

Solution on page 77

The Tightrope

No circus act is complete without a tightrope walker. How does he keep his balance and why does he use a long bar?

Solution on page 83.

Coasting Home

The other day I was looking at an atlas of the world and came across a table that gave the lengths of coastlines of different islands. After thinking about it for a little while, I realized that this was not very informative. Why not?

Solution on page 76.

Lingering in the Rain

While out walking the other day I got caught in a sudden rain shower and made a dash for home. Just as I got to my front door the rain stopped as suddenly as it had started.

While I dried out, I wondered if I would have been just as wet if I had not bothered to run. What do you think?

Solution on page 76–77.

A Bolt from the Blue

A report in the newspaper the other day told about a farmer who had had a lucky escape. He was in a field when a near-by tree was struck by lightning. At the time he happened to be standing next to a cow that promptly fell down dead.

Had the cow been suffering from a weak heart, or is there another more scientific reason?

Solution on page 79.

Around the Bend

A friend of mine lives due south of London and works due north. The M25 is the highway circling London. Every day my friend traveled up on the M25 to work by going to the east of London, and in the evening, for a change of scenery, he would travel back to the west. However, he realized that by doing this the same two tires were on the outside on both the outward and return journeys and must have traveled farther. They must have been wearing faster. He changed his behavior by traveling both to and from work by the easterly route.

Thinking about the situation in more detail, he realized that the lanes of the highway in one direction were outside of the lanes traveling in the other direction. So the differential wear on the tires, although less than before, would still exist. Was my friend correct in all his reasoning?

Solution on page 77.

Fishy Business?

I have been watching my goldfish swimming around in its bowl, and have noticed that occasionally it swims to the surface to take a gulp of air. I know the effect would be very small, but wondered if this would cause the bowl to increase or decrease in weight, or stay the same.

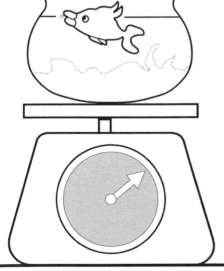

While thinking about it, I wondered if I could simulate this effect by using an underwater syringe to draw in air from the surface like a kind of "mechanical fish."

What do you think?

Solution on page 78–79.

High Flyers

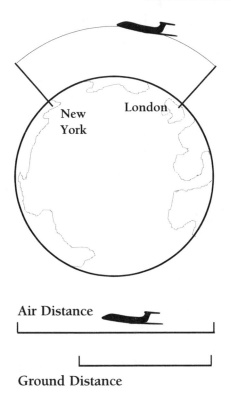

New York
London

Air Distance

Ground Distance

We know that as the radius of a circle increases, so does its circumference. This must mean that the higher an airplane flies, the greater the distance it has to fly to its destination.

On a flight between London and New York, an airliner has to fly an extra two miles because of its high altitude.

Forgetting air-space congestion as a possible argument, why do airliners not fly by a shorter, lower route?

Solution on page 78.

Maximum Speed

Cars have a maximum speed. If a driver keeps the accelerator pressed down, the speed increases up to that maximum. While a rocket fires its engine in outer space its speed will increase, no matter how fast it is already going.

Why does a car have a maximum speed, but a rocket does not?

Solution on page 75.

Once Upon a Time

A long time ago two captains of sailing ships had a wager with each other as to which was the faster way to sail around the world: east to west or west to east. They decided to put it to the test by having a race. So one day at exactly the same time, they set sail from a small island each going in the opposite direction.

Some months later they happened to arrive back at the same island at exactly the same time. They were just about to decide that the bet was off when they compared logs and found that there was a discrepancy of two days between them.

Is it possible to account for the discrepancy and decide who circumnavigated the world faster?

Solution on page 78.

Inclined Not to Move?

If a car is parked facing downslope, a marble placed on the floor will roll towards the front. If a car accelerates along a horizontal road, a marble will roll towards the rear of the car.

Would it be possible to accelerate a car down a slope at such a rate that a marble would remain stationary on the floor?

Solution on page 79.

And the Earth Moved

Seismographs are sensitive instruments for detecting earthquakes. They consist of a large mass suspended by springs. When an earthquake happens, the disturbance travels through the Earth, making the instrument vibrate. However, the mass tends not to move because of its inertia. This difference in movement is amplified and written out as a trace.

Can you explain why most places in the world receive two traces for each single earthquake event? Can you also explain why a few places detect only one?

Solution on page 79.

Bursting Barrels

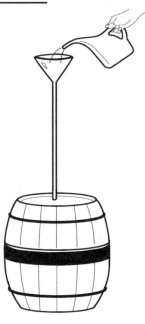

There used to be a trick popular in Victorian times. A beer barrel was completely filled with water. A long thin tube had been attached to the top of the barrel, which was empty at the start of the trick. By pouring a very small amount of water into this tube the barrel could be made to burst, demonstrating that small causes can have large dramatic effects.

Suppose one jug of water was needed for the trick. If a narrower tube with half the cross-sectional area had been used, how much water would have been needed?

Solution on page 80.

The Stalling Plane

If a plane stalls, the pilot will be unable to pull out of the dive immediately, but is likely to succeed after having nose-dived for some time. Why?

Solution on page 81.

Snow Drift

Have you noticed how much more snow is proportionally deposited on the sides of posts and poles rather than on the sides of buildings? Why?

Solution on page 82.

The Thirty-Inch Ruler

Hold a ruler horizontally, balanced across your index fingers.

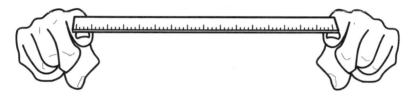

Then try to move your fingers together so that both slide. It does not work. The ruler first slides over one finger and then over the other, but never both together. Do you know why?

Solution on page 84.

The Target Practice

You are standing on top of a high-rise building in Greenwich with a long-range, high-powered rifle, perfectly aligned. You are trying to hit a telegraph pole in Louth, Lincolnshire. Assume that you can see the target through the telescopic sight on your rifle. Would you aim straight at the target, to the left, or to the right of it?

To give you a clue, both Greenwich and Louth are on the 0° meridian.

Solution on page 82.

Perils of Diving

A great deal of research has gone into diving and the physiological aspect of ascent to the surface. Suppose, using scuba equipment at a depth of 25 meters, you find that the equipment is faulty, forcing you into a speedy ascent, hoping that the air in your lungs is sufficient to take you to the surface.

Would you in such an emergency gradually release air as you ascend, thus reducing your air reserves, or would you hold your breath?

Solution on page 81.

The Eskimo

There are many stories of shipwrecked sailors dying of thirst or being driven insane by drinking seawater. Eskimos have no source of freshwater. Does polar ice contain salt, and if so how are the Eskimos dealing with the problem?

Solution on page 81.

Mountain Time

Before climbing the Matterhorn you synchronize your very accurate spring-driven watch with your friend's at the base camp. You want to beat the record to the top and you have arranged to record the time, by radio contact with the base, the moment you touch the flagpole at the summit.

Strangely enough, you cannot agree on the time, although there is nothing wrong with either watch. A difference, although small, still persists after you return to base. Can you explain why this should be.

Solution on page 80.

The Returning Boomerang

The boomerang is a missile used mainly by Australian aborigines. Originally used as a weapon, tossing the boomerang has recently become a sport. It is generally held vertically in the right hand, although some are designed for left-hand use. The fascinating aspect of this weapon is its ability to return to the hands of the thrower.
  Can you explain this phenomenon?

Solution on page 80.

The Channel Tunnel

When the French and English shafts broke through to join the tunnel, the workers and local dignitaries celebrated the event by opening several bottles of the best champagne. To their disappointment the bottles hardly popped and the poured champagne was flat. Yet when the celebrities returned to the surface they felt sick and could not stop burping. Why?

Solution on page 81.

Wartime Experience

Those of us in Great Britain who lived through World War II will remember, without much nostalgia, V-1 and V-2 rockets replacing the bombing raids that had become too expensive for the Germans in terms of losses in men and planes.
  It is in man's nature to be adaptable. We soon found that the V-1s presented no danger as long as you could hear them. However, as soon as the buzzing of the engine stopped, it was advisable to dive under the bed. No such strategy was available with the V-2. Why not?

Solution on page 84.

45

The Flag and the Weather Vane

Assume that the wind blows with uniform force from the same direction. A weather vane will remain stationary while a flag, even if perfectly smooth and fully spread by hand to start with, will flap once released. Why?

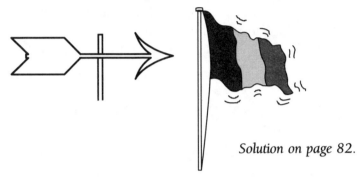

Solution on page 82.

Steering the Boat

Can you steer a boat on a lake if it is entirely becalmed and you have no oars? The answer must be no!

Suppose the circumstances are the same except that your boat is drifting in a fast-flowing river. Can you then steer with the rudder?

Solution on page 84.

Flight of Birds

Did you ever ask yourself what it is that enables birds to fly? After all, they are heavier than air and therefore do not float.

To fly they would need lift and forward propulsion. How do birds achieve this?

Solution on page 86.

The Concorde

Even if you have never flown in the Concorde, that ingenious product of Anglo-French technology, you know that this plane is the only supersonic aircraft in commercial use. You also know that Mach 1 stands for the speed of sound.

Now answer the following questions:

1. Do you hear the sonic boom when, traveling in the Concorde, you reach Mach 1?

2. Is the Concorde's ground speed always the same when the plane breaks the sound barrier?

Solution on page 55.

The Air Tube

There is the story of an escapee from an island prison who evaded recapture by swimming underwater, using a metal pipe he had stolen from the prison workshop as an air tube.

Sharks apart, is the story feasible, and is there a limit to the depth at which the swimmer can breathe through the tube?

Solution on page 84.

The Hottest Place on Earth

Death Valley is a depressed desert region in southeastern California. Most of the valley is below sea level and it has the distinction of being the hottest place in the world. In 1913 a temperature of 134°F was recorded—the highest ever.

Elementary physics has taught us that hot air rises and cold air sinks. Would you therefore not have expected Death Valley to be a cool or moderately warm place, particularly as it is almost entirely enclosed by mountain ranges?

Solution on page 87.

The Perforated Water Can

Drill a number of small holes in a can and fill it with water.

1. Fit an airtight lid and the water stops leaking from the holes. Why?

2. Drill a hole in the lid and the water now flows. Why?

3. Remove the lid, run your finger through the leaking streams, and they, if the holes are not too far apart, will merge as shown, forming a single stream even after you have removed your finger. Why?

Solution on page 83.

The Panama Canal

The construction of the Panama Canal is regarded as one of the greatest technical achievements of all time. It was completed ahead of schedule and was in full operation by the summer of 1914. With the map of the world in mind, one would have no doubt that the Canal runs from west to east. Surprisingly, the Pacific end lies somewhat east of the Atlantic end.

There are some other interesting aspects to which the reader is invited to find an answer:

1. At the last lock, as the gate is opened, any ship will move out to sea without tugs and without using its own power. What makes it move?

2. One would assume that the water levels in the Atlantic and Pacific are the same. However, there is a difference at times of as much as 12 inches. Why are the ocean levels not the same?

Solution on page 71.

TONGUE-IN-CHEEK
PHYSICS

This section includes ideas that might be theoretically feasible but, due to real-world limitations, may not be realized. Other items are fallacies, or contain flaws in reasoning that are only too obvious to the intelligent reader. As such, these problems are more amusing than intellectually challenging.

The Ultimate Weapon

D. Stone, in a letter to *Geotimes* (1969), describes a geophysical weapon which, if used by a nation with a large population, could have a devastating effect on an enemy.

Suppose, for instance, the residents of China, who now number nearly one billion, were to all jump in unison on pogo sticks. Such an action could set up shock waves which would amplify with each jump, reaching perhaps a magnitude of 5 on the Richter scale. This could, for example, destroy parts of the United States. To protect itself the USA might organize jumps timed to cancel the offensive waves. But, considering the difference in populations, the Americans would either need to jump from a greater height or call on the NATO allies for jump assistance.

After seriously considering all possibilities, and acting on the advice of their defense experts, history would suggest that the Pentagon would decide in favor of countering the threat in a manner other than by pogo stick. However, would this Chinese super-weapon work?

Solution on page 85.

Air Mail?

When posting a parcel, you have to pay more if the parcel is heavier. There is a scale of charges linked to the weight of the parcel. The post office is quite right to charge according to the mass of the parcel. However, I have noticed that the machines used are spring balances, which measure weight rather than mass.

This means that I could employ the following scheme. When I wanted to send an object through the post, I could enclose it in an over-sized box, which would leave space for me to put a helium balloon in with it. The parcel has the extra mass of the helium and the fabric of the balloon, but because of the upthrust on the balloon the parcel would weigh less, saving me postage.

Presumably, if the balloon were large enough, the parcel would have negative weight, and the post office would have to pay me to transport it. Would the scheme work?

What machine would correctly measure the mass of the parcel?

Solution on page 89.

Charging for a Flight?

I know from my school physics classes that whenever a conductor moves through a magnetic field, a voltage is produced across it. This was discovered by Michael Faraday in 1831 and is the basis for all electric generators.

It occurs to me that most aircraft are made of metal, and they move through the Earth's magnetic field. I did a calculation to see how much electricity could be generated. Unfortunately, the Earth's magnetic field is very weak, so even the Concorde would generate only about a quarter of one volt between its wingtips. But this might be enough to run some small electronic devices without the need for batteries or generators. Is this scheme feasible?

Solution on page 88.

The Spying Game

In war as in industrial espionage, the main problem is to avoid detection when transmitting messages. Microdots and invisible ink are old hat, and new methods are devised all the time.

Paul Goodwin was employed on a top-secret project concerning isotopes and his problem was to transmit a message to his principals abroad. Coded messages were too dangerous and therefore unthinkable. After much consideration he devised what he considered to be a foolproof method.

There are 26 letters in the alphabet. As any message would also contain punctuation marks and blanks (between words), we are dealing with a total of about 40 symbols. Paul assigned a two-digit number to each, starting with 01 for "A", 02 for "B", and so on until he reached 38 for a full stop and 39 for an interval between words. (ISOTOPE MASS would, for instance, read: 09 19 15 20 15 16 05 39 13 01 19 19.) With the help of his powerful computer, Paul Goodwin translated the message into one long number.

The reader is now asked to permit some intellectual license. By putting a decimal point before the number, we can consider the whole code to be expressed as the length in meters of a bar made of precious metal (to avoid any chemical reaction with the atmosphere). This bar would have a length of more than 9 but less than 10 cm.

Let us assume that Paul has the means to cut the bar to the required length, which would require fantastic precision.

Let us also assume that the recipient of the bar, having

equally precise measuring equipment, would obtain the same decimal number and decode the secret message.

Assuming that the precision needed for this operation were available, would it work, at least in theory?

If not, can you think of a procedure which would make it theoretically viable?

Solution on page 87.

Perpetual Motion II

A weight is suspended over a pulley and attached to the ground by a helical spring. As the world turns under the moon, its gravity makes the weight move up, turning the pulley. The pulley is connected to a generator, creating electricity.

Would the setup work? Is it a perpetual-motion machine? Where does the energy come from?

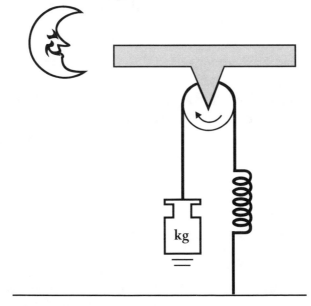

Solution on page 83.

Height and Weight

Let us suppose that there is no limit as to how tall a person could grow. Now imagine someone growing rapidly, becoming progressively taller and therefore heavier.

The question the reader is asked to consider is this: Would the person's weight increase forever?

Solution on page 86.

The Time Machine

Most readers will have heard of the International Date Line (IDL), established as an irregular line, drawn by convention through the Pacific Ocean, substantially along the 180th meridian. IDL marks the place where the date changes.

Crossing the line from west to east, travelers gain a day and, conversely, lose a day traveling east to west.

The position of the IDL has been arbitrarily designated and is partly curved to accommodate eastern Siberia, then bulges westward again in order to avoid crossing land.

Let us now assume that a successor of the Concorde can take its passengers around the equator, circling the globe in six hours. Flying west to east it will cross the IDL four times in 24 hours and consequently lose four days. Starting, say, on the 18th of March it will land on the 14th of March. Continuing the journey the passengers will go back into history, eventually reliving the birth of Christ. Traveling east to west they will travel into the future, giving them a decisive advantage in the lottery on one's return.

Is there something wrong with this reasoning?

Solution on page 88.

Time Bomb

Here is a design for a bomb that requires no explosives. Imagine that the bomb was made of a material that was a perfect reflector of sound. Inside it has a spherical cavity. At the center of this cavity sits a ticking watch. The sound inside will build up, as it can not escape. Eventually the sound energy would be so great that it could no longer be contained. At this point the bomb would explode.

Would this design work?

Solution on page 89.

SOLUTIONS

LED Lights

If you filmed the LED with a high-speed camera to slow down the effect, you would see it flash red and green. But our eyes, by a process called "persistence of vision," work much more slowly and you would be unable to see different-colored flashes.

Also, our eyes do not have a color receptor for each color in the rainbow. There are only three types of color receptor in our eyes, sensitive to mainly red, mainly green, and mainly blue. The other colors are detected by triggering more than one receptor to a greater or lesser extent. Color televisions make use of this fact; if you look closely at the screen you will see that the picture comprises only red, green and blue dots. Our flashing LED will stimulate the red and the green sensor simultaneously, giving the sensation of yellow light.

The Concorde

1. There is a common misconception that a sonic boom occurs only at the moment when a plane exceeds the speed of sound. In fact, any aircraft traveling faster than sound creates a large pressure disturbance that travels along with the plane. If this pressure disturbance passes a person on the ground it is detected as an explosive sound. That is why the Concorde is not allowed to fly supersonically over land. As this pressure wave does not pass the passengers inside the plane, they are unaware of the sonic boom.

2. No. "Breaking the sound barrier" is a common expression meaning traveling faster than sound. Sound does not always travel at the same speed; it is affected by the air temperature. As air is colder at high altitudes, the speed of sound is correspondingly lower by about half a meter (20 inches) per degree of Celsius. At 0°C, the speed of sound is about 328 meters (1075 ft) per second, and at 16°C about 338 meters (1108 ft) per second.

Black to the Future

White is, strictly speaking, not a color but the presence of all colors. A white bulb gives out all visible colors. Black, also, is not a color but the complete absence of all light. By definition, therefore, there can be no such thing as a source of "black light."

Car Headlights

It would work. However, it was not introduced because the disadvantages would outweigh the benefits.

1. Not being able to see the headlights of approaching cars would be dangerous, particularly when visibility is reduced during heavy rain or fog.

2. The windshield's polarization would absorb a fair amount of light in the street scene, reducing visibility generally.

3. At any rate, it would be difficult to provide an effective filter for windshields in view of their irregular shapes.

4. Colored stress patterns would become visible in the windshield and this would be distracting for the driver.

The Safety Lamp

The explosive gases would still infiltrate through the mesh and ignite inside the lamp. The screen would quickly conduct away the heat, thereby preventing the flame from escaping to cause an explosion. The change in the flame also gives a warning of danger to the miner.

The Setting Sun

A number of books have been written on the subject, Minnhert's *Light and Color in the Open Air* (Dover, 1954) to mention one.

In simple terms, the vivid coloring of sunset is due to the scattering of sunlight by atmospheric dust, whereby the blue light of the spectrum is scattered more than the longer wavelengths of the red end. Dust from many sources, for instance volcanoes, contributes to the variety of color display.

For once, I would suggest that you forget physics and continue to enjoy the unique experience of a glorious sunset.

Lo-Fi?

Even with correctly connected speakers you find that in most parts of the room there will be several frequencies of sound that combine destructively while others combine constructively. This means that some frequencies will be quieter than they should be, while others are louder; but this is not normally noticeable among the thousands of other frequencies present. At any point equidistant from each speaker, however, you can guarantee that all the frequencies will combine constructively, giving the best listening position.

But if you wire one of the speakers with the reverse polarity, at the central position all frequencies combine destructively, giving the worst possible listening situation.

Frequent Changing

There would be even worse problems if two or more transmitters attempted to broadcast the same program on the same frequency.

As radio is a type of wave, at some points the two waves from the two transmitters could just happen to be synchronized so the radio would pick up a very strong signal. At other places the waves could be completely out of synch and would cancel each other out. Thus, as you moved around, the radio would continually go from extremely loud to no signal at all within the distance of a few meters.

The Foehn

There are three questions to be answered:
1. Why is the wind warm? The meteorological conditions have to be right for the cold mountain air to sink in the form of a fast-moving wind. As the air sinks it encounters the higher pressures found at lower altitudes and is compressed. This compression causes warming; the result of a phenomenon known as adiabatic process (see Glossary).
2. Why is the wind dry? The air was originally cold and cold air can hold less moisture than hot air.
3. Why does it have psychological effects? Nobody is sure that it does, or if so what the reason might be.

Sound Reasoning

Although light and sound are both wave forms, there are some significant differences, the major one being wavelengths. Sound waves are typically one meter in length, whereas the wavelength of light is about one ten-thousandth of a millimeter. Also, sound is "coherent" (continuous) whereas all light sources (except lasers) produce light in the form of millions of small packets which are all jumbled up.

1. Waves will curl around obstacles (diffract) when their size is of the same order as the wavelength. Thus, sound will easily diffract around buildings, trees, etc. This can also be seen for light, but the obstacles have to be much smaller. For instance, a diffraction pattern can often be seen while looking at a bright point of light through a closely woven material.

2. Light can also be made to combine destructively, just like sound, but because of the small wavelength and lack of coherence this can be difficult to observe. This is what causes the characteristic "speckled" appearance of a spot of laser light and the colors seen in soap bubbles and films of oil.

Perpetual Motion I

If the magnet were strong enough to draw the ball up the slope, it would be powerful enough to make it jump across the hole. The ball would therefore stick to the flange or magnet.

Throughout history there have been attempts to create a perpetual motion machine: something that would run forever, producing energy. It was not until people had an understanding of the law of conservation of energy that they realized that it was an impossible quest.

Energy can exist in many forms (heat, light, chemical, electrical, potential, kinetic, etc.). The law of conservation of energy says that energy cannot be created or destroyed; all you can do is convert one type into another (e.g., a car converts chemical energy into kinetic, heat, electrical energy, and so on). There is as much energy in the universe now as there was millions of years ago. This law is so fundamental that the British Patent Office refuses any application based on a device which, once in motion, will produce external work indefinitely without any supply of energy.

Although perpetual motion can be observed on a sub-atomic level, for instance electrons orbiting a nucleus, no energy can be drawn from such a system. A perpetual motion machine is, sadly, therefore impossible.

Getting Cold Feet

The two floors were probably at exactly the same temperature. However, carpet is a much poorer conductor of heat than are the tiles. As the tiles were taking heat from my feet faster, the tiling *felt* colder than the carpet.

Breath on Your Hand

There are two possible explanations involving: a) evaporation and b) expansion.

The blown air moves more rapidly, causing moisture in the skin to evaporate quickly, which produces a cooling effect. As the air passes through pursed lips, it expands and cools. This is the opposite effect to air being compressed in a bicycle pump becoming warm.

You can prove that both effects are present with a piece of clear food-wrap. If you place this over your hand, preventing evaporation, the blown air still feels slightly cold. Wetting the outside of the food wrap allows evaporation to occur, restoring the full cooling effect.

In the Dark?

There are two reasons why solar eclipses are so rare:

1. The moon's orbit is not in the same plane as the Earth's orbit around the sun. This means that every month when the moon is on the "right side" it is usually too far to the north or to the south to block the sun's light.

2. If the two orbits were in the same plane, a total eclipse would occur every month, but it would still be seen from very few places on Earth. This is because the moon's shadow, when cast on the Earth, is only a few miles across. Only if observers happen to be in this shadow area will they experience a total eclipse.

The Drinking Bird

The construction of this gadget is as complex as it is ingenious.

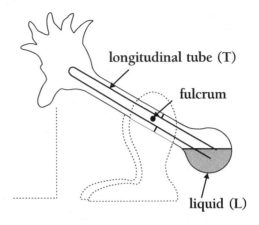

longitudinal tube (T)

fulcrum

liquid (L)

The base is partly filled with a suitable vapor-forming liquid (L) and connected to the head by the tube (T). As T is immersed in L, it produces vapor in the base as well as in the tube and the head. Because the immersion of T in L forms a seal, the two vapor systems are not connected. One should also mention that the center of gravity (COG) of the contraption is below the fulcrum, so that its position is, to begin with, stable, as illustrated.

Now you wet the top of the bird. As water evaporates, the vapor in the head cools, lowering the pressure. The higher pressure of vapor in the base forces liquid up the tube, relocating the COG until the bird's head tilts forward, dunking its head in the water. In doing so, T is lifted out of L and the two pockets of vapor connect in the tube and equalize pressure. This, together with the buoyancy created by immersing the head, moves the COG again to below the fulcrum. The bird rights itself and the whole cycle begins again.

What makes this toy interesting is in considering where the energy comes from to keep it going. The answer is that the energy comes from the heat in the water—which gets progressively colder.

We are used to the idea that useful energy comes in very concentrated forms, for example gas or coal. It is unusual to find a machine that can exploit low-grade energy. If perfected, such a machine could be used to power ships, which would then not need to carry fuel. Seawater could be pumped on board, its heat energy extracted to propel the ship, and any resulting cold water/ice dumped back into the sea.

Ringing Tone

There was a strip of railing nearby. Sound reflected back from each individual support rail arrived as a series of small echoes separated by a small interval of time. This caused a distinct ringing tone, the frequency of which was determined by the distance between the rails.

The Light Fantastic

In one of the rooms, the light is provided by sodium lights, which provide a very pure yellow light. In the other room the light is provided by a mixture of red and green lights. Red and green lights, when mixed, appear yellow. In the first room, both red and green paint appear black, while in the second room the red and green paints reflect their own colors.

Squaring the Circle?

The apparent gravity increases the farther one moves from the point of rotation. The gravitational effect would therefore be greatest at the edges of the room. For example, a marble placed in the center of the floor would roll towards one of the two edges, which are parallel to the axis of rotation.

Even though the floor is perfectly flat, it would feel as though there were a hump in the room, that one would be walking "uphill" towards its center.

In a Spin?

The clock gave the answer. Sundials were invented in the northern hemisphere, the shadows indicating the time as it moves around the dial.

Clocks were invented as a mechanical equivalent of the sundial, the hands representing the shadow. The shadow cast by the sun must therefore move clockwise. This means that, looking down on the North Pole, the Earth must rotate anti-clockwise, i.e., west to east. Consequently the sun rises in the east and American time lags behind European time.

The Toy Boat

As the steam enters the cooler tubes it condenses, creating a vacuum that will draw in water. However, such water as is sucked in will not do so in the form of a jet, but will come from all directions and will not therefore have the same propulsive force.

The Rainbow

Part of the mysterious beauty of the rainbow:

1. You can see the spectrum only if the angle of refraction between the sun, the drop of water, and your line of vision is between 40° and 42°.

2. Lunar rainbows are also possible but rare, because moonlight is not as strong as sunlight and its intensity varies with the phases of the moon.

3. This phenomenon is called dew-bow and is caused by the water drops on the grass.

Gravity

There is no way to distinguish between 1) gravity and 2) acceleration. However, in a rotating spacecraft "gravity" will appear greater farther away from the center of rotation, so there are several things that he could do:

1. Weighing an object with a spring balance will give a smaller reading near the ceiling.

2. A marble placed in the center of the floor will roll towards the edge.

3. Two plumb lines will diverge rather than hang parallel.

4. He would not be able to spin a coin.

Shape of the Flame

The characteristic shape of a candle flame is caused by convection as the hot exhaust gases are pushed up by the colder, fresher air.

In weightless conditions there would be no convection as there is no up and down. If there were no drafts the flame would be spherical, but would not last very long because local oxygen would soon be used up, with no air currents to replenish it.

Vision and Sound

With two ears you can identify the direction from which a sound is coming. The same principle is used in direction-finding equipment to locate the source of illegal radio transmissions.

With only one eye you would be unable to see objects in three dimensions, nor would you be able to estimate distances accurately.

Information Relay

The time it takes for an object to orbit the sun is determined only by its distance therefrom. The Earth takes 365.3 days to orbit the sun because it is 93 million miles from it. At a position halfway to the sun, the period of rotation can be calculated as 130 days. The satellite could not orbit the sun and maintain a movement in conjunction with the period of orbit of the Earth.

The Spy Satellite

The orbit of a satellite must be around the center of the Earth, which is the center of its gravitational field. There is no way in which a satellite can be synchronous with Baghdad's latitude. It follows, therefore, that a satellite can hover only over a point on the equator.

The Satellite

At first the atmospheric drag is very small and has the effect of making the satellite fall into a lower orbit. The speed in orbit is determined by height: the lower the orbit the faster the speed. The satellite therefore travels faster.

This may seem like a puzzling contradiction to the law of conservation of energy. But the total energy of the satellite, potential and kinetic, has in fact decreased by the amount of heat produced by friction.

Eventually, as the orbit spirals downward, the frictional drag will be great enough to slow down the satellite as it falls into the main body of the atmosphere. The increase in orbital speed is similar to the effect on a skater who pulls his outstretched arms nearer to his body, thereby increasing his speed of rotation.

The Fourth Dimension

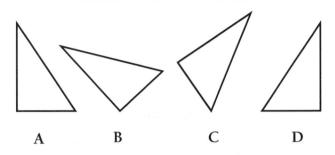

A   B   C   D

The illustration above shows four right-angled triangles with all their sides and angles in common. They are, therefore, clearly congruent. But our 2-D friends, looking at the four triangles, are faced with a baffling phenomenon.

They can see that, by moving them around on the plane on which they presently lie, A, B, and C can be superimposed. They are directly congruent. D, however, cannot, despite meeting the test of congruency with the other three, because it is oppositely congruent. No amount of shuffling around on the plane on which it lies will bring it into coincidence with the other three. The only way to achieve that is to pick D up, turn it around and then re-introduce it into the two-dimensional world. If our 2-D friends could only see that—if their imagination is up to it—then they can conceive of a 3-D world.

Do we encounter similar situations in our world? Yes: mirror images of irregular geometrical bodies. Take, for instance, a left and right hand (ignoring the minor differences which make them not quite perfect mirror images in practice). By definition they are not directly congruent, and cannot be superimposed. Presumably a fourth dimension could do the trick.

Pull of the Sun

Let us consider a simpler system first. Suppose there were only the Earth and the sun. If the Earth were stationary in space, the two would be drawn together by their mutual gravitational attraction. However, there is a more stable possibility: the sun's gravitational force is used to maintain the Earth's orbit rather than drawing it closer.

The same process explains why the moon orbits the Earth. The sun's gravity is used to maintain the Earth/moon system in orbit. If all orbiting suddenly ceased, all the planets and their moons would instead fall towards the sun.

One Side of the Moon

No, it is not a coincidence. It is well known that the moon exerts tidal forces on the Earth's oceans. What is less well known is that there is also a tidal effect on the Earth's solid surface. The total result of this is to slow the rotation of the Earth, each day being slightly longer than the preceding one. This is why occasionally a "leap second" has to be added to the length of a day to keep clocks accurate.

It is assumed that at some time in the past the moon was rotating faster, and that tidal forces in its surface slowed its rotation until it became synchronous. This phenomenon is quite common in the solar system; for instance, Titan takes 15 days and 23 hours to orbit Saturn and to rotate once about its own axis.

Some cosmologists take a different view. P. Goldreich, in an article in *Scientific American*, believes that the synchronous rotation of the moon is caused by the fact that the moon's mass is not uniformly distributed and that the Earth's gravitational field acts upon the heavier half of the moon.

Faster Than Light?

My friend correctly identified three situations in which something travels faster than light. However, none of them contradicts Einstein, whose work could be more correctly summarized as: "The maximum speed at which information can travel is represented by the speed of light in a vacuum."

In the first two examples, nothing physical moves faster than light. The barber's pole stripes only appear to move along the pole. The spot of light at one end of the sweep is different from the one at the other end as it is made up of different photons. Neither system could be used to carry information.

The third example has been observed experimentally. Such particles break the "light barrier" and produce minute photonic flashes equivalent to sonic booms.

Upon Reflection

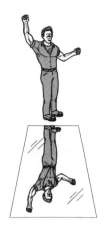

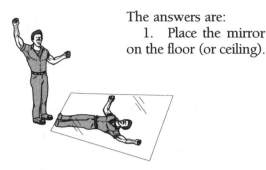

The answers are:
1. Place the mirror on the floor (or ceiling).

2. Place the mirror at a 45° angle to the floor.
3. I cannot think of one either.

Round or Flat?

As long ago as 340 BC the Greek philosopher Aristotle argued that the Earth was a round sphere. This he based on two observations.

He recognized that the eclipse of the moon occurred when the Earth came between sun and moon. Had the Earth been a flat disc, the shadow would at times be elliptical or a straight line. However, the Earth shadow on the moon was always round.

The other observation concerns ships coming over the horizon. Had the Earth been flat, you would expect the whole ship to become visible as it approached. What in fact can be seen are, first, the masts, and then the hull, rising on the horizon.

Insulated from the Truth

Notice that the total heat loss is 100%. This is always so, however well insulated the house is; all the heat put in will eventually get out until the temperatures inside and outside are equal. What is important is the *rate* at which heat escapes. Doubling the insulation of the house will halve the rate of heat loss, and thus also halve the rate at which heat will have to be supplied to maintain the same temperature. This would also halve the energy bills.

But the proportion of heat loss through the walls, windows, and roof would remain roughly the same.

The Olbers Paradox

Olbers himself provided the counter-argument.

1. While light from some of the galaxies we observe left them millions of years ago, the light of many millions of stars has not yet reached us. Our own galaxy is more than one hundred thousand light-years across.

2. The paradox assumes that all or most stars in the observable universe are alight at the same time. This is a misconception, because the life span of a star is limited to about 10^{10} years. Although this is a long time, it is still finite.

3. As we have seen in "The Fourth Dimension," the universe is probably not infinite.

4. Light may also be absorbed by dust.

So Near and Yet So Far

If a telescope (or a pair of binoculars) is being used in the correct way, it is said to be in "normal adjustment." When utilized in this way, the eye is perfectly relaxed when using it. When a perfect eye is relaxed, it focuses at infinity. A telescope does not, therefore, give you an image that is closer. It does, however, make the object look bigger.

A magnifying glass increases the focusing power of the eye, allowing the object to be placed closer. There are many circumstances in which it works by not magnifying at all. Try this simple experiment to see what I mean. Hold a piece of writing so close to the eye that it cannot be seen clearly because it is blurred. Now insert a magnifying glass between the eye and the writing. You should now find that the writing is clear enough to read but is no larger than it was.

Time to Leave?

Thousands of years of evolution have ensured that trees shed their leaves at the optimum time. If they drop their leaves too early, then valuable time would be lost. If the leaves fall too late, they could be damaged by frost, and this would be detrimental to the plant as a whole.

Glaringly Obvious?

Strangely enough, the answer to this question is "Yes." If there are just two crossed polarizing filters, no light will pass through the combination. If you insert an extra filter between the other two, at 45 degrees to them, then some light will travel through this triple combination. In fact, the amount of light transmitted through the triple combination is about half of what would pass through a single filter.

This is because some light will always manage to get through a double combination at 45 degrees to each other. So some light will pass through the first and second filter. Some light will also pass through the second and third filter. Therefore, some light will pass through the complete triple combination.

A Balancing Act

You might think that the correct result is the average, 122 grams, but this would be true only for a balance with equal arms. This balance must have unequal arms, and the formula for this is the square root of the product: $100 \times 144 = 120$ grams

Proof: If the scale's balance when empty is inaccurate, the two arms must be of different lengths (a, b).

Let the corresponding readings for the object X be W1 and W2

$$X \xrightarrow{\quad a \quad b \quad} W1 \qquad\qquad W2 \xrightarrow{\quad a \quad b \quad} X$$

At balance:
$$aX = bW1 \rightarrow a/b = W1/X \text{ and}$$
$$aW2 = bX \rightarrow a/b = X/W2$$

Therefore:
$$W1/X = X/W2 \text{ or}$$
$$X = SqRt \ (W1 \times W2)$$

Substituting:
$$X = SqRt \ (144 \times 100)$$
$$= 120 \text{ grams}$$

A Moon Mystery

A dark part of the moon, which is not directly illuminated by the sun, nevertheless receives some sunlight indirectly, which is reflected from the Earth's surface.

A Quiet Cup of Tea

Where the water is being heated it is locally hotter than elsewhere. Small bubbles of water vapor will form, which collapse as they move to cooler parts. The "singing" sound is caused by these thousands of tiny collapsing bubbles. When the water is close to boiling, the bubbles survive, causing the kettle to go quiet.

The Expert Driver

Do not lock the brakes, for there is less friction between the tires and the road if they are sliding rather than rolling. Under normal road conditions you would require about 25 percent less distance to stop if you kept the tires rolling.

It is very difficult to judge exactly how much pressure to apply to the brakes to maximize deceleration while avoiding locking. Many modern cars have anti-lock braking systems that detect when the wheels are about to lock and automatically momentarily release the brakes.

The Toy Balloons

It is a surprising scientific fact that as a balloon gets larger its internal pressure drops. It will therefore be more difficult to expand the smaller balloon. A very simple experiment can demonstrate the fact. Try gently blowing up two identical balloons simultaneously, one partially pre-inflated. You will find that the larger one will inflate in preference to the smaller one.

The Hot Water Tap

The answer is so simple that I should have known the first time I noticed. The hot water heats the tap valve first, which expands, reducing the flow. After the whole tap has warmed up, the effect disappears.

Color of the Colorless

There is no substance that is perfectly transparent. Light is gradually absorbed as it passes through an increasing thickness of any substance. That is why it is completely dark at the bottom of the ocean.

If all colors are absorbed equally, there is no resultant color. Ordinary glass has a slight tendency to absorb magenta, leaving the green hue seen through thick glass.

The sky appears blue because that color has the shortest wavelength and it is scattered by small dust particles in the air.

There can be two reasons why mountain lakes look blue. First, the water is reflecting light from the sky. Second, water from melted glaciers is often colored because of the high mineral content.

The Spray Gun

The air traveling through the blow tube at considerable speed reduces the air pressure in the conduit tube. The liquid in the container is subject to atmospheric pressure and is therefore forced up, and out of, the conduit tube in the form of a fine spray.

Soap Bubbles

The air pressure inside the bubble is greater than the atmospheric pressure outside. It could not be less or it would collapse.

The surface tension necessary to form the bubble is directed towards the center, trying to reduce the size, and works therefore in tandem with the atmosphere. The inside pressure must, therefore, be greater to balance the two forces acting against it.

The Well-Trained Sunglasses

The sunglasses we are referring to contain crystals of silver bromide, which are photosensitive and undergo a change if exposed to light. The silver ions are transformed into silver atoms which darken the lenses. The functional usefulness of this phenomenon is its reversibility. As soon as the light intensity is reduced, the silver atoms recombine to form silver bromide.

The Bicycle Pump

The resultant temperature changes are caused by a phenomenon known as adiabatic process. In addition to an explanation, a brief definition of what is termed an adiabatic process is in order (see Glossary).

Such a process is one in which no heat (at least in the short term) is entering or leaving a system. If gas is compressed in a cylinder, for instance a bicycle pump, it undergoes an adiabatic change and heats up. If gas is released, as in the compressed air bottle, the expansion results in the cooling of the gas.

The Panama Canal

The two responses desired are:

1. The canal is fed by a number of freshwater lakes including Gatún and Miraflores. When the last gate leading to the ocean is opened, the freshwater level will still be higher than the denser saltwater, creating a flow to equalize levels and providing a drift for the vessel.

2. The salinity of the Pacific is higher than that of the Atlantic, and therefore denser, which accounts for the lower level of the Pacific.

Cool It!

It is a mistake to believe that by simply opening the refrigerator door you will achieve your objective. You might momentarily feel cooler, but you will soon increase the temperature as the motor releases more heat than it reduces in the refrigerator.

The only way to succeed is to switch the motor off and open the refrigerator door. The cold air will mix with the warm air and achieve an overall reduction.

The Hourglass

The weight of the hourglass will not change, although some of the grains of sand will be weightless, being in free-fall. This will be balanced by the impact force when the grains hit the bottom. There may be some slight fluctuating at the beginning and end of the process, however.

Golf Balls

In fact, dimpled golf balls travel about four times farther.

The effect is caused by the backspin imparted to the golf ball by the club. As the top of the ball spins backwards, it drags air which would otherwise have traveled beneath the ball. This air has to speed up to travel the extra distance, and this causes lift in a very similar way to an aircraft wing. If the ball were smooth, the spin would have no effect.

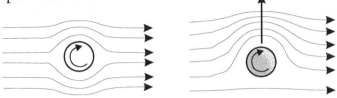

Crafty Hover

Attach an overlong piece of string to the balloon and hold the free end. The string will then form a curve, traveling down from the hand to a low point and then rising up to the balloon. Cut the string at its lowest point. The balloon will then be left with exactly enough string to cause it to hover.

What Makes Glue Stick?

Molecules of a substance attract each other by a process called cohesion. Molecules will also attract molecules of a different substance, and this process is called adhesion. The trick is to find substances whose molecules are very adhesive.

To work effectively the molecules have to be in close proximity, which is why adhesives are usually liquid and the surfaces involved should be as clean as possible.

Rather surprisingly, water is quite a good adhesive. If, for example, you wetted two pieces of wood, placed them together and put them in a freezer, once frozen the pieces of wood would be very difficult to part.

Theoretically, if one could hone two surfaces to a degree that molecular contact was possible, no adhesive would be needed. In practice such surfaces are contaminated by dust and other impurities, so that adhesives are needed.

Grounding the Automobile

The metal strip is designed to ground the vehicle. Rubber is a good insulator, and it is possible for a car to build up a charge of static electricity.

There are several reasons why people might want to use such a device. Some worry that this static charge could cause a spark which could ignite gasoline vapor, causing a fire. It is possible that the strip would avoid such a spark, but it is by no means proven that there is a fire risk in the first place.

Other people use it to avoid an electric shock as they close the car door on leaving. This static buildup is caused by the rubbing of their clothes against the fabric of the seats. The metal strip is unlikely to solve this problem; not wearing rubber-soled shoes would probably be more effective.

Some people have a theory that travel sickness is caused by static electricity. Any improvement in the situation here is likely to be more psychological than real.

Water Level

As to the classic question, the reader probably knows that the water level drops simply because in the boat the brick displaces water equivalent to its weight, while at the bottom of the pool it displaces only its volume of water.

Now to the second question. As the boat *begins* to leak, surprisingly the water level in the pool remains constant. This is because two effects exactly cancel each other out. As some water leaks into the boat, the pool's water level tends to drop. However, the boat is now heavier, displacing more of the pool's water, causing the level to rise.

As long as the boat remains buoyant, the pool's water level remains constant. As soon as the boat becomes submerged it displaces less water, causing the level to fall.

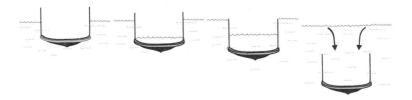

Insulating Metal Pipes

The pipe will lose heat by three processes: conduction, convection, and radiation.

The amount of heat lost through radiation depends on the excess temperature of the surface and the total surface area. The insulation will reduce the surface temperature, although it also increases the surface area. As we have seen, the insulator may well increase the heat lost through conduction.

However, most of the heat is probably lost through convection. The amount of this will be determined by the surface temperature, which is greatly reduced by the insulating material. If there were no insulation, and we just relied on using a thicker pipe, this would have very little effect on the surface temperature but would increase the surface area. This would increase the heat lost through convection and radiation.

Helicopter Flight

As the helicopter moves forward and the rotor blades rotate, there is a difference in the lift provided by the forward-moving and backward-moving rotors. This is because the velocity of the aircraft is added to the forward-moving rotors and subtracted from the backward-moving rotors. The faster the helicopter moves, the worse the effect becomes; if the forward motion equaled the backward speed, there would be no lift at all on that side.

Increasing the speed of rotation of the rotors would help, but there is a limit to this. The tip of a rotor is traveling the fastest, and if it should break the sound barrier locally, the resultant disruption to the airflow could cause the helicopter to crash.

Not the Half of It!

For symmetrical objects the bisectors will be coincident, but this is not true for asymmetrical objects.

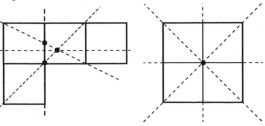

74

My Head in the Clouds

As the wind flowed over the mountain, it rose and then descended. This wave continued for a few oscillations over the plain. The conditions were such that, at the top of the wave, the air cooled enough for the water vapor to condense, forming a cloud at that point. As the air continued along its path, it sank and warmed sufficiently for the water droplets to evaporate again. This caused a series of clouds which were being continuously created at the back and destroyed at the front. The clouds were therefore stationary although the air was moving.

Long Live the Evolution

It is unlikely, since these diseases do not manifest themselves until after an individual ceases to be sexually active. Therefore, someone who had a resistance to one of them would not have any advantage over someone who did not, so his genes would be just as likely to be passed on as anyone else's.

Dropping a Bombshell

When a bomb is released, it has two motions. It starts to fall vertically downward. It is also moving forward, initially at the same speed as the plane, although friction gradually slows it down.

If the plane is flying at a low level, there is a danger that the bomb is still directly beneath the plane when it strikes the ground and explodes. This could damage the plane. The function of the parachute is to rapidly retard the horizontal velocity of the bomb, so there is less danger of it inadvertently damaging the low-level bomber.

Maximum Speed

The car stops accelerating when the thrust of its engine equals the drag of resistance or friction. A rocket in outer space experiences no friction, so there is no limit to its acceleration.

There is, however, a maximum speed at which anything can travel; so the rocket will never exceed the speed of light.

Coasting Home

When measuring a coastline, the result you achieve will depend entirely on how accurately you attempt to make the measurements. For instance, suppose we start taking our measurements at one end of a large bay. Should we measure just the distance straight across to the other side of the bay? This would give us an undermeasurement; so perhaps we had better measure around the bay.

But as we go around, we find more smaller bays and inlets; should we measure across or around them? If we measure around them, we will find even smaller indentations. Every time we decide to measure inside a new indentation, the measured distance increases. If we went down to molecular levels, this first bay alone would give us an almost infinite distance!

Hotheaded

I had dried my ears with a towel and therefore they felt the real temperature of the air. My hair was still wet, so the fast-moving air was rapidly evaporating the water. Evaporation produces cooling (this is the principle used in the domestic refrigerator), and therefore the air felt cool on the wet hair.

Lingering in the Rain

Like lots of "real life" problems, this is more complicated than it first appears. So let's make some simplifying assumptions:

1. There was no wind—the rain was falling vertically.
2. I was wearing a hat—so we are considering the rain that I walked into, not the amount falling on my head.
3. The rain was falling at a steady rate.

Now let's think about the space that I moved through; we could imagine an invisible me-shaped tunnel stretching from where I am standing to my front door. Rain is continually entering this tunnel from the top, and leaving at the bottom. As it is raining steadily, there is as much rain entering the tunnel as leaving, so an equivalent system would have the raindrops hovering stationary. It is now obvious that however quickly I move though my personal tunnel, I will collect exactly the same number of raindrops.

But remember, the rain stopped as I reached my doorstep, so if I had not run, I would have got through only about half my tunnel before the rain disappeared from the other half. By running I had actually become wetter than if I had continued my leisurely walk.

If the rain had not stopped but had continued falling steadily, I would have been just as wet either way. However, there is an element of chance; if it had started raining more heavily, I would have got wetter by walking.

Suppose I had not been wearing a hat. The top of my head continues to get wet as long as I am in the rain. Running keeps my head drier.

If there had been a head wind, this would have made me wetter the longer I stayed out, i.e., continuing to walk would have made me wetter. If there had been a tail wind, running would have decreased the wetness to a point where the speed of the running equaled the wind speed, when my body would not have got wet at all!

So, is it better to run? It depends....

A Question of Stability

a) a cone
b) a sphere
c) a dumbbell
d) a sphere with weight embedded off-center

If the only equilibrium a particular object has is unstable, with no stable or neutral position, a small perturbation would have a most dramatic effect. It would never be able to attain another equilibrium position and would keep moving forever, searching for one.

Around the Bend

My friend was correct in assuming that if a car travels in a circle the outside tires travel farther and therefore wear more than those on the inside.

However, it is a surprising fact that the extra distance traveled by the outside wheel is not affected by the radius of the circle around which the car is traveling. So his new way of driving to and from work would succeed in equalizing the wear on the tires.

Sheep May Safely Graze

The evolutionary process can be much more subtle than one realizes. Unlike most other plants, grass grows from the base of its stalk and not its tip; it can therefore cope with grazing fairly successfully. If all the grazing animals were eliminated, other plants would then be much more successful while grass, being a small plant, would lose out to those other plants in the competition for light.

You just have to think of environments where there are large numbers of grazing animals to realize how successful this strategy is for grass.

High Flyers

Even taking the extra distance into account, it is still cheaper to fly at a higher altitude where the aircraft is flying above the turbulent weather patterns. Also, the thinner air presents less drag, meaning less fuel is consumed. Finally, it may be possible to take advantage of the fast winds found at that higher altitude.

Once Upon a Time

If they both set off at the same instant and also returned at the same time, then both their journeys would have taken the same time. However, the two ships would have counted different numbers of passing days. If the island had experienced the passage of "X" days, the ship traveling west against the rotation of the Earth would have counted X – 1 days because it would have made one less rotation than the world. The ship traveling with the spin of the globe would have counted X + 1 days.

This anomaly was corrected by the adoption of the International Date Line.

Fishy Business?

Let us consider the syringe first. Drawing some air into the syringe will increase the weight of the system because of the extra weight of the air. But at the same time the piston in the syringe moves back. The water level rises slightly, increasing the volume of the system, thus increasing the upthrust exerted on it by the atmosphere. By the Archimedes Principle, the increase in upthrust will equal the weight of the extra air. So there will be no overall change in weight.

The same would be true for the fish, so long as the drawn air was not compressed or absorbed. As both those things are likely to happen inside the fish, the change in upthrust will be less than the increase in weight. Therefore, the overall weight of the system will increase.

A Bolt from the Blue

When the lightning bolt strikes a tree, the electricity continues into the ground causing a voltage gradient across the surface. Because the cow's front legs are more than a meter from its back legs, there can be a large voltage difference which drives a high, and in this case fatal, current through the poor beast. Because humans have only two feet close together, there was not enough voltage to cause the farmer too much trouble.

Inclined Not to Move?

Assuming that the marble is frictionless, it will always roll down the slope with the same acceleration relative to the slope. If the car is accelerating at a rate less than the marble's, the marble will roll towards the front. If the car's acceleration is greater than the marble's, the marble will roll towards the back of the car. When the two accelerations are exactly the same, the marble will be stationary relative to the car.

And the Earth Moved

There are two ways in which earthquake waves can travel around the world. First, they can travel across the Earth's surface, rather like waves travel across the sea. Second, they can travel directly through the center of the Earth. These waves travel at different speeds and have different distances to travel, so they arrive at the seismograph at different times.

The Earth does not have a uniform density. The denser core acts like a lens, focusing the vibrations at some points while leaving other places in a "shadow."

Ring Around?

If a relatively weak material were used, for example string, the ring would collapse to the ground. If the material were very rigid, it would be in unstable equilibrium, so would touch the ground at one point.

The ring could be made stable if it were rotating, but air friction would soon bring it to rest.

Bursting Barrels

The trick works because the water pressure at one point depends on depth, not the weight of water above. If the tube connected to the top of the barrel is narrow, a small amount of water can cause a large increase of pressure in the barrel by increasing its depth below the surface.

If a narrower tube were used, then even less water would be needed to increase the depth. If the tube were half the area of cross section, then half the amount of water would be needed to attain the same height in the tube. So half the jug would be sufficient.

The Returning Boomerang

The return boomerang has a length of 30 to 75 centimeters, curving to the left, and capable of more than 90-meter throws. There have been several attempts at explanations. T.L. Mitchell in 1846 suggested that it was caused by the skew combined with the spinning motion. This does no more than beg the question. A more convincing explanation is offered by F. Hess in "The Aerodynamics of Boomerangs" (*Scientific American*, Nov. 1968).

The boomerang is an airfoil and therefore subject to a lift, which is greater on the top half because it is turning in the same direction as the boomerang, whereas the bottom half is turning in the opposite direction.

Mountain Time

As you ascend from sea level the atmosphere becomes thinner and offers less resistance to the spring, making your watch run faster.

The Eskimo

Polar ice does indeed contain salt. Such ice, melted down, is as undrinkable as seawater. However, over time, the brine in the ice blocks will migrate downward, because of gravity. This draining effect will make the melted ice drinkable after about a year, and it will be almost completely free of salt after several years.

The problem does not affect all polar regions, as some ice is formed by precipitation. However, many areas are arid, with insignificant snowfall, and such little as there is will be blown away by winds of up to 100 mph.

Among the various desalination techniques, freezing has been developed as an alternative method, based on the different freezing points of fresh- and seawater, but the equipment needed is beyond the reach of Eskimo communities.

The Stalling Plane

A plane can stall when it is traveling too slowly for its wings to provide sufficient lift. When that happens it starts to fall. If, as with most planes, there is more weight at the front, it will fall nose first. As it picks up speed, air flows over the wings, once more providing lift. Eventually there will be enough lift to enable the pilot to pull out of the dive.

This also accounts for the characteristic looping flight that you can see with a paper airplane.

Perils of Diving

As one ascends to the surface of the water, the external pressure on the body decreases. The lungs contain air that expands and could cause rupturing. Divers are instructed that, during an emergency rise to the surface, they should leave their mouth open, allowing the expanding air to escape.

The Channel Tunnel

At the bottom of the shafts the atmospheric pressure was much greater, so that more of the carbon dioxide remained in solution to be released at the lower pressure on the surface.

The Target Practice

You should aim to the left of the target. Any moving object will be deflected to the right north of the equator, and to the left south of the equator, in relation to the rotation of the Earth.

This phenomenon is called the "Coriolis effect," named after the French physicist Gaspard de Coriolis (1792–1843), who first analyzed it mathematically. The Coriolis effect is of great importance to meteorologists, navigators, and the military (see Glossary).

Snow Drift

The wind driving the snow diverges many meters in front of a large building, thus dispersing the snow before it hits the windside face. A smaller object does not divert the wind, permitting the snow to build up.

What Color Is the Sun?

I can think of five answers:

1. **white** – almost by definition, the color of sunlight is white.

2. **all colors** – one could argue that there is no such "color" as white. Isaac Newton discovered that white light is really a mixture of all visible colors.

3. **yellow** – the most abundant color in sunlight.

4. **black** – the sun appears white because it is emitting light, rather in the way a light bulb looks white while it is on; the bulb element is gray when the bulb is off. But what color would the sun be if it cooled down? Scientists know that the sun is close to being what is called a Black Body Radiator. So if it cooled down it would be black.

5. **no color at all** – black is not really a color but is the absence of all colors.

The Flag and the Weather Vane

Again the Bernoulli effect takes over. No flag is perfectly smooth. The tiniest imperfection will make the airflow speed up as it crosses the ripple, reducing air pressure on the ripple side, increasing it on the other side, making the flag flap.

The Tightrope

Walking a tightrope is possible only if you keep your center of gravity precisely above the rope. This is a bit like keeping upright on a stationary bicycle. The balancing pole is deliberately heavy, with weights placed in the tips. It makes things easier in two different ways. The bar has a large inertia, and by moving it sideways the walker can adjust his position. Also because the bar is heavy at the ends and bends downward, this has the effect of lowering the center of gravity of the walker. If this could be lowered to below the rope, then the walker, although still looking very precarious, would in fact be very stable.

The Perforated Water Can

And the three answers are:

1. The pressure in the can is less than the pressure outside. The inside pressure is caused by a few centimeters of water while the outside pressure is caused by the atmosphere, which is equivalent to about 10 meters of water. Water will not flow from a low- to a high-pressure area, as that would be like flowing uphill.

2. By drilling a hole in the top of the arm, the atmospheric pressure is applied to the surface of the water; the pressure outside is still atmospheric, but the pressure inside is now atmospheric plus a few centimeters of water. The water therefore flows out.

3. Surface tension prevents the water streams from separating again. Surface tension is a phenomenon by which the molecules at the surface of a liquid are held tightly by the cohesive forces of the molecules beneath. Water has quite a high surface tension.

Perpetual Motion II

In theory the system would work; the change in weight, however, is very small (about one-thousandth of one percent), so the amount of energy is minute. Still, this is where the tide's energy comes from.

The energy comes from the rotational energy of the moon. The tides have the effect of increasing the moon's orbital radius. This is not, therefore, a perpetual motion machine.

Steering the Boat

The rudder will have an effect only if there is relative motion between the boat and the water. This answer begs the question "Is there such relative motion?" Most likely yes. However, there are so many forces acting upon the boat that this question cannot be answered with absolute certainty.

First, there is gravity. Imagine the river to be frozen. Ignoring friction, the boat would slide down the slope. Then there is buoyancy, resulting in a component force downriver, partly counteracted by drag. Air resistance adds to the drag, while a wind upstream or downstream has an effect. Furthermore, the river flows at different rates in the middle and near the banks.

It is theoretically possible that all these forces combine to synchronize the motions of the boat and river for a limited period, which would make steering impossible.

The Thirty-Inch Ruler

The finger that slides first is the one with less friction between it and the ruler. The friction depends on the weight of the ruler on the finger.

Try as you might, you can never balance it precisely. A minute difference is sufficient to favor one finger against the other. However, as the sliding finger approaches the middle, the weight on it increases; therefore, it will stop and the other finger will begin to slide.

Wartime Experience

The V-2 rockets exceeded the speed of sound and therefore you could not hear their approach. You could hear only the detonation after they hit the target, too late to take evasive measures.

The Air Tube

The story is perfectly feasible, provided the swimmer keeps at least one meter below the surface. Any deeper and the water pressure would prevent chest expansion to enable the lungs to inhale.

The Ultimate Weapon

It would not work, even theoretically. In the first instance the major devastation would occur in the country resorting to such a weapon although, according to David Stone's theory, there would be damage in the USA provided the jumps were timed in regular intervals to amplify the resulting ground waves. Stone estimates, however, that a jump would have to take place about *once every hour* for some considerable time. So, apart from the devastation at home, the attacking country's economy would certainly suffer irreparable damage.

Theoretically at least, the USA might indeed consider inviting all NATO members to organize counter-jumps, carefully timed, to cancel the ground waves initiated by their adversary, as an alternative to using advanced weaponry in defense of their shores and avoid the possibility of starting a world conflagration. Before they do so, however, let us now do a simple calculation to work out how much energy would be produced by each attacking jump, and the extent of the danger to the USA.

Assumptions:
No. of Chinese	= one billion
Average mass of 1 person	= 50 kg
Pogoing height	= 0.5 meters
Acceleration due to gravity	= 10 ms^{-2}

Calculation:
Energy produced:
$$= 1{,}000{,}000{,}000 \times 50 \times 10 \times 0.5$$
$$= 2.5 \times 10^{11} \text{ joules of energy}$$

This would produce a local earthquake of about 5.3 on the Richter scale, causing trees to sway and some damage locally. Put another way, this is about one four-hundreth the amount of energy released by the atomic bomb over Hiroshima.

The theory goes that the waves of energy from the jump would spread around the world and focus on a point directly opposite. However, the Earth would rapidly dissipate the energy, converting it into heat. We know this because, when an earthquake occurs at one point in the Earth, it does not cause damage at its antipodes. So the USA and NATO can stand down their anti-pogo defenses.

Flight of Birds

Lift is easily explained; it is the same principle as applied to aircraft, involving the Bernoulli effect (see Glossary).

Look at a typical wing, be it that of a plane or a bird:

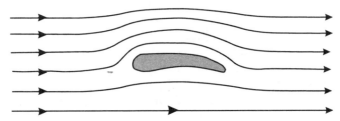

The air above the wing has to cover a greater distance and therefore travels faster, reducing the pressure. There is greater pressure below the wing; both effects produce lift. This enables the bird to glide and soar through the air.

It is more difficult to explain what provides the forward propulsion. High-speed photography shows that a bird's wing changes shape, flexing in a complex fashion, providing a backward thrust to the air during part of its cycle. As a reaction to this the bird is pushed forward, in a way analogous to a swimmer using the breast stroke.

Certainly a straightforward flapping would be ineffective. Neither do I accept Storer's explanation that the twirling of birds' feathers acts as a propeller.

Height and Weight

Yes and no. Our weight is conveyed by the gravitational attraction of the Earth. However, part of this force is counterbalanced by the centripetal force due to the rotation of the globe. This force is zero at the poles, increasing to a maximum at the equator. A person standing, for example, in Quito will have his center of gravity move up as he grows, thereby increasing the centripetal force, while at the same time reducing the effect of gravity. Consequently, a point will come when the resultant weight will start to decrease. When the center of gravity is about 22,000 miles (36,000km) high, the person would be weightless and efficiently in orbit. Needless to say, at one of the poles, the weight would keep on increasing, irrespective of height.

The Hottest Place on Earth

A number of factors contribute to these extreme conditions. The mountain range on the west side rising to more than 3,300 meters absorbs the moisture of the west winds, which, on descending east of the Rockies, are adiabatically heated and dried, turning the valley into a hot, arid desert.

The Spying Game

No, the operation would not work theoretically. A minute variation in temperature would completely demolish the code, due to thermal expansion.

You could cope with the problem of thermal expansion by converting the decimal into a fraction and cutting the bar into two pieces, one to represent the numerator and the other the denominator. This would be purely a mathematical solution, inasmuch as thermal expansion would affect both bars, leaving the ratio, and therefore the fraction, unaffected.

On a physical level it would not even work theoretically. The problem is the tremendously small distances that one is soon involved in. Every letter added to the message would involve a hundredfold increase in accuracy. Sending messages with only two letters (e.g., OK) would probably be feasible, for this would require an accuracy of one-tenth of a millimeter.

To send a five-letter message (e.g., ATOMS) would require an accuracy of one-tenth millimeter, which is about half the diameter of an atom of iron. As you cannot have half an atom, the sending of such a message by this method is impossible.

Some readers might think that lengthening the bar would provide a viable solution. Ignoring the logistic problem connected with sending a long bar, let us re-examine the position.

Remember that to send the message ISOTOPE MASS, we have to transmit the number 0919152015160539130111919. Let us suppose we send a bar whose length is this number of millimeters. How long is the bar?

$$91{,}915{,}201{,}516{,}053{,}913{,}011{,}919 \text{ millimeters} = 10^{23} \text{ mm}$$
$$= 10^{17} \text{ km*}$$

* 600 million times farther than the distance (93 million miles) between the Earth and the sun.

Charging for a Flight?

No. It is true that a voltage would be generated between the wing-tips as the plane flew through the Earth's magnetic field. However, to make use of it we would have to complete the circuit, and that would require running wires to the ends of the wings. Those wires would also be traveling through the Earth's field and would also generate the same voltage. It would be like connecting two identical batteries back to back: the two effects would cancel each other out.

If we consider other forms of transport, we might be more successful. By a similar process a train would generate a voltage which could be picked up from the rails by contacts lying beside the track. I calculated that a train traveling at 100 mph (160 kmph) would generate about 1.5 millivolts. That is obviously not enough to solve any energy crisis!

NASA is experimenting to see if this is a viable power source for spacecraft. A long wire, several miles long is trailed through the Earth's magnetic field. The electric charge is dissipated from sharp points at both ends of the wire.

The Time Machine

Yes, there certainly is. First, let us distinguish between natural time and an arbitrary time zone. A time zone is an area within which everyone has agreed to use the same time. However, every point on the Earth's surface has its own natural time. As one travels east, the natural time is farther ahead, 1 hour for every 15 degrees of longitude.

Let us suppose that the aircraft travels eastwards around the world so fast that it would complete the journey in one second, and it started from a point just to the east of the International Date Line. As it traveled it would encounter a natural time which was progressively further ahead.

When it arrived at a point just to the west of the International Date Line the time would be 24 hours ahead. This extra day is then lost as the date line is crossed. The final time is then exactly one second after the start.

One cannot, therefore, travel into the past or future by using the International Date Line.

Air Mail?

In theory the scheme would work, although I doubt that any post office would pay you if they had to deliver a floating parcel. It is commonly stated that spring balances measure weight (i.e., the force of gravity) and that beam balances measure mass (i.e., the amount of substance in an object). This is not strictly true; the beam balance compares weights, and therefore would not be affected by a change in gravity. For example, kitchen scales would give the same reading on the moon, but a spring balance would not.

In this case, however, even a beam balance would not give a correct reading of the mass of the parcel. I do not think that there is a common instrument that would. Either the parcel should be weighed in a vacuum or a machine would have to be invented that measured the inertial mass by, perhaps, gently oscillating the parcel.

Driving on Ice

Good drivers know:

1. To increase friction where it matters you will use the blanket in front for a front-wheel-drive vehicle, otherwise in the rear.

2. The lower the torque the better; therefore, you should start in second gear at slow speed.

3. To stop skidding you should turn your front wheels into the skid.

Raw or Cooked

If you spin the egg it will stand on end if it is hard-boiled, similar to a top. A raw egg is unstable because it is asymmetric and the contents are viscous and therefore will not spin.

Time Bomb

The design would not work. First, there is no such thing as a perfect reflector of sound. Second, the amount of energy in a watch is quite small. But even without these two objections the proposal would still be impractical.

Air is a very insufficient transmitter of sound and consequently sound energy is continually being converted into heat. The energy thus escapes from the box in the form of heat.

GLOSSARY

Assuming that readers may not be familiar with all the scientific terms used in this book, it should be useful, for ease of reference, to define here some concepts not part of everyday vocabulary.

absolute zero — the temperature at which the molecules of any object are stationary; this, therefore, represents the coldest possible temperature.

adhesion/cohesion — adhesion is the force of attraction between two different types of molecule. The force of attraction between similar molecules is called cohesion.

adiabatic process — a term used in thermodynamics, referring to a condition in which no heat enters or leaves the system, although pressure and volume are varied. For example, the air in a bi-cycle pump will heat up if compressed, because no significant heat transfer will take place immediately. Another example, with the reverse effect, is an aerosol can. On releasing the contents, the temperature of the can will drop. Other systems in daily use, such as automobile engines and refrigerators, exhibit adiabatic phenomena.

Archimedes' principle — states that when any object is immersed in a fluid there is an upthrust that acts on the object which equals the weight of the displaced fluid.

Bernoulli effect — discovered and formulated by the Swiss mathematician Daniel Bernoulli in 1738. The principle states that, as the speed of a moving liquid or gas increases, the pressure within that fluid decreases. This principle is an important aspect of aerodynamics and covers the flow over surfaces, such as the wings of aircraft and ship's propellers. As the air flows over the upper surface of a wing it speeds up, and consequently suffers a reduction in pressure as compared to the lower surface. The resulting difference in pressure provides lift to the aircraft.

center of gravity — the point in any object where all the weight could be considered to be concentrated without affecting the properties of the object. Such an assumption can greatly simplify calculations.

cohesion — *See* **adhesion**.

conduction of heat — the process by which heat energy can travel through a substance by the transfer of vibrational energy between adjacent molecules. This is greatest in solids.

convection of heat — the process by which hot fluid is forced to rise by being displaced by colder fluid.

Coriolis effect — a deflection caused by the rotation of the Earth.

diffraction — a process by which a wave curls around small objects or spreads out as a result of passing through a small gap.

drag — *See* **friction**.

energy conservation — the scientific law that states that energy (heat, light, sound, electricity, etc.) cannot be created or destroyed, but can only be converted into another form.

evolution — the theory proposed by Charles Darwin that states that species change over long periods of time, as a result of which individuals that possess advantageous differences are more successful in procreation.

fluid — a substance that flows: a liquid or a gas.

friction — a force that results from one object moving relative to another with which it is in contact.

gravity — a force of attraction between any two objects. The force is very small and is only noticeable if at least one of the objects is very large.

LED — stands for Light Emitting Diode, a small solid-state device that produces light, without heat, when a small electrical current is passed through it.

lift — the upward force produced by the flow of air over wings. If the lift disappears because of a disruption in this flow, the wing is said to stall.

light-year — the distance that light travels in a vacuum in one year, about 5.88 trillion miles (or 9,460,000,000,000,000 meters).

mass — measure of how much material there is in an object. This is independent of the force of gravity (weight) acting on the object. Weight is measured by a spring balance and mass by a beam balance.

Newton's laws of motion — (1) An object will remain at rest or in uniform motion unless acted upon by an unbalanced force. (2) When an unbalanced force acts upon an object, it will accelerate at a rate proportional to the force and inversely proportional to the mass of the object. (3) For any action on an object, there will be an equal and opposite reaction.

perpetual motion machine — refers to a mechanical device doing work and operating perpetually without any supply of energy, other than that which is generated by the device itself. Such a system cannot exist, as it is contrary to a well-established physical law, namely the principle of "conversion of energy." This law is so fundamental that patent offices have been known to refuse any application based on perpetual motion.

Perpetual motion is perfectly feasible in the absence of friction; for example, electrons rotating around a nucleus or an object flying through space. However, as soon as energy is extracted from such a system, it will slow down, destroying its perpetual motion. Needless to say, any device using the force of gravity or variation of atmospheric pressure or temperature would not qualify.

polarization of light — light is a waveform that has an electric and magnetic vector. In all forms of natural light these vectors are not aligned. If aligned, the light is said to be polarized.

polarizing filter — A filter that allows light to pass which is vibrating in one direction only. *See* **polarization of light**.

principle of flotation — states that a floating object displaces its own weight of fluid.

radiation of heat — the process by which heat can travel through a vacuum or a fluid. It is a form of light (infrared).

Rayleigh scattering — named after Lord Rayleigh (1842–1919), a type of deflection of electromagnetic radiation by particles in the matter through which it passes. The radiation photons bounce off the atoms and molecules without any change of energy (*elastic* scattering), changing phase but not frequency, as opposed to *inelastic* scattering.

refraction — the process by which the direction of light changes when it travels from one medium to another of different density. This is how lenses work.

stall — *See* lift.

surface tension — a condition existing on the surface of liquids giving them film-like characteristics. The tension is explained as resulting from intermolecular forces of the liquid. Examples: A water beetle can ride the surface of water; the near-perfect sphere of a soap bubble; a small quantity of mercury poured onto a plane assumes a near-spherical shape, flattened only slightly by gravity.

weight — *See* mass.

INDEX

Page key: puzzle, **solution**

ABOUT THE AUTHORS

ERWIN BRECHER was born in Budapest and studied mathematics, physics, psychology, and engineering in Vienna, Czechoslovakia, and London. He joined the Czech army in 1937. After the Nazi occupation of the Sudetenland, he escaped to England via Switzerland. Engaged in aircraft design during the war, he later entered the banking profession, from which he retired in 1984. Writing his first book manuscript in 1992, Erwin Brecher saw it published in 1994. Since then, he has written ten more books of puzzles and on scientific subjects for U.S. and European publishers, including one in German. In 1995, he was awarded the "Order of Merit in Gold" from the city of Vienna, in recognition of his literary achievements.

A recent work, *The IQ Booster*, published both in the USA and UK in 1996, expounds a novel approach to IQ testing. Readers' reply cards indicated an improvement in IQ rating of up to 23 points.

Erwin Brecher is a regular contributor to magazines and radio. Currently, he devotes his time to playing bridge and chess, and to several new literary projects.

A member of Mensa, Erwin Brecher and his wife, Ellen, have made their home in London, England.

MIKE GERRARD joined Mensa while still at school and founded the Norfolk group in 1963. He won the *Evening Standard* "Brain of London" competition in 1980. Having studied physics and education at London University, Mike has been a physics teacher, a Head of Resources, a Head of Science, and a senior teacher in comprehensive schools.

In 1987 Mike received a Master of Science degree in information technology at Kingston Polytechnic (now University) and joined International Computers Ltd. as an Education Consultant. He subsequently joined Computer Associates, where he currently works.

Mike has compiled many puzzles used in specialist magazines, was contributor to *The Holiday Book of Things to Do* (1973) and *The Second Book of Things to Do* (1975), also by Knight. He has also written for the Schools Council, a former advisory body in the UK.